PLAYING GOD

Genetic Engineering and the Manipulation of Life

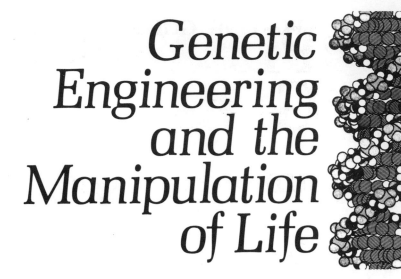

PLAYING GOD

June Goodfield

RANDOM HOUSE
NEW YORK

Grateful acknowledgment is made to the following for
permission to reprint previously published material:

Harcourt Brace Jovanovich, Inc.: Four lines from
"Gerontion" in *Collected Poems 1909–1962* by T. S. Eliot.
Copyright 1936 by Harcourt Brace Jovanovich, Inc.
Copyright © 1963, 1964 by T. S. Eliot.

Macmillan (Journals) Ltd.: Chart beginning on page 210,
reprinted from *Nature*, Vol. 262, July 1, 1976, page 3.
Prepared by Mr. Colin Norman based on the DNA
recombinant research guidelines issued by Dr.
Frederickson, Director of the National Institutes of
Health.

Library of Congress Cataloging in Publication Data

Goodfield, G   June.
Playing God.

Includes index.
1. Genetic engineering—Social aspects.   I. Title.
QH442.G66   174'.2   77-6023
ISBN 0-394-40692-3

Manufactured in the United States of America

9 8 7 6 5 4 3 2

FIRST EDITION

For
Peter and Frances
Ramsbotham

I have a feeling that if your structure (of the DNA molecule) is true . . . then all hell will break loose, and theoretical biology will enter a most tumultuous phase.

—*Max Delbruck: Letter to Francis Crick and James Watson*

I would in fact say it is quite different than, and probably much more important than, the development of atomic energy . . . Modern biology is leading us to the point where we have the ability to manipulate genes, and manipulate the essence of life, if you wish, of various organisms under various conditions. And the change that (it) will bring to our perception of ourselves and our ability to control ourselves, I think, will outweigh atomic energy, which brings on the one hand . . . a certain fear of atomic explosion, and on the other hand, an alternate form of energy. (That) is . . . just really an increase in style, but not really an increase in a qualitative sense in our lives.

—*Dr. David Baltimore, Nobel Laureate, testifying to the U.S. Senate Subcommittee on Health, September 2, 1976, on recombinant DNA*

Nature does not need to be legislated. But playing God does.

—*Spoken by a Middle European microbiologist at the Asilomar Conference, February 1975*

PREFACE

Over two years have passed since the Asilomar Conference, that unique event in the history of science when scientists themselves called for certain restrictions on certain types of experiments. But the genesis of this book is much older, the Asilomar Conference and its aftermath providing for me both a focus for ideas that had been quietly gestating for some time and an occasion to write these down. In February 1972, at a conference on science and social values at the University of South Florida, Tampa, I argued that the existing social contract between science and society was rapidly dissolving and in the process of hammering out a new one, certain cherished assumptions and hallowed beliefs would be swept away. (This argument permeates the book, but the substance is in Chapter 5.) The advent of the technique of recombinant DNA, with its wide-ranging implications, has, in my opinion, merely accelerated a process that had already begun. But it has at the same time provoked a whole range of new questions and opened up a spectrum of issues and debates. During the eighteen months that it has taken to research and write this book, I could see one question follow another and one set of concerns displace a former set, so that the whole process came to resemble the forces passing down a freight train just as it begins to move. A tug at one end—in this case a new simple scientific technique—produced an effect which, point by point, was transmitted all the way down the line, till the problem of the relationship of the scientific profession to contemporary society was itself pulled into the arena of public scrutiny. So though this book deals in great part with the science of recombinant DNA and genetic engineering (Part I), it does so only insofar as they have a bearing on problems of the wider social context (Part II). I believe it is in this wider context that the whole issue should be considered. I hope that those who read this book will come to understand the science and why scientists regard it as so exciting and important. Yet at the same time I hope that scientists as well as the public will come to appreciate the wider issues and see why it is important for the layman to understand and participate in the wider debate.

This book inevitably is an interim report on a debate that may well become even more intense in the next few years. I have tried to give a picture of where we presently stand scientifically, legally and morally with regard to recombinant DNA and genetic engineering. The full history must await publications from the Technology Studies Program of Massachusetts Institute of Technology, where the history of the whole episode is being painstakingly recorded, orally and by a wide variety of documentation, by Charles Weiner and Rae Johnson. Those who wish to delve more deeply into the technical aspects of the matter can consult the issue of *Science* magazine for April 8, 1977, which carries a series of articles about the science and an editorial by Dr. Maxine Singer. Similarly, those who wish to read at first hand the views of the proponents and opponents, as well as those of politicians, economists and others, should read the proceedings of the forum held by the National Academy of Sciences, which is to be published in the fall of 1977 as a book, *Research with Recombinant DNA.*

Once again I must emphasize—as I did in my book *The Siege of Cancer*—that much of this writing depends on the ideas and work of others. My debt to them is enormous. I am grateful to those who have already written about these issues; to the scientists at Asilomar who participated in the crucial discussions and whose views I was able to hear through the courtesy of the National Academy of Sciences of America, which holds the tapes in its archives. I have, besides, personally interviewed many of them in America, Britain, Switzerland and France, and their names will be found in these pages. I deeply appreciate the time they gave me and the patient courtesy with which they taught me. I am especially grateful to Drs. Paul Berg, Sydney Brenner, David Jackson, Kenneth Murray, Maxine Singer, Charles Weissmann and Norton Zinder, all of whom tolerated extended interviews and often a whole battery of follow-up telephone calls.

But my greatest debt is to two people, Drs. Peter Carlson and Shaw Livermore, who, respectively, guided me through the maze of recombinant DNA (Chapters 2 and 3), and the moral and humanistic issues (Chapter 9). Scientist and historian, each in turn drew me into their professional and humane concerns with such openness and warmth that I now feel I have known them for years, such was the empathy they demonstrated both for my job and the problems at hand.

The administration of the University of Michigan, Ann Arbor, in the persons of Vice-President Frank Rhodes and Dean John Gronvall gave me every opportunity and personal help. This enabled me to be present and cover much of the story of the first open debate on recombinant DNA, as it developed on their campus.

Finally, many friends gave me encouragement, advice and support ranging from criticism to discussion to monitoring the papers and journals so that I was kept up-to-date: Joy Cull, Joan Thornton, Maria de Sousa, Tony Smithyman, Delphine Parrott, Francis Bennett, Hilary Rubinstein, Neil Nyren.

Lastly I must—with deep affection—thank my mother. The three months of 1976 that it took to write the first draft of this book coincided with the hottest summer on record in Europe, and because of the drought all garden hoses were banned. So after each morning's writing, I found myself carrying buckets and buckets of water to keep alive a most beautiful garden and an even more valuable vegetable patch. As I carried and dug and hoed and weeded and picked, she, too, at seventy-eight, hoed and watered and encouraged and teased. In other words, she kept me sane. I could not have done it without her.

J.G.

CONTENTS

Preface   xi

I: THE SCIENCE AND THE SCIENTISTS

1: *Creating New Problems*   3 •
2: *Creating New Molecules*   12
3: *Creating New Species*   31
4: *Creating New Men?*   46 •

II: THE SCIENTISTS AND SOCIETY

5: *Creating New Issues*   77
6: *Creating New Commandments*   91 •
7: *Creating New Guidelines*   113
8: *Creating New Precedents: The Law and Politics*   139
9: *Creating New Moralities*   161

Conclusion: Walking the Tightrope   190 •
Afterword   202

Appendix I: *The NIH Guidelines*   210
Appendix II: *Extracts from the Guidelines for the Use of Recombinant DNA Molecule Technology in the City of Cambridge, Massachusetts*   213

I
THE
SCIENCE
AND THE
SCIENTISTS

1
Creating
New Problems

Science comforts, Art disturbs.

—*Attributed to Georges Braque*

The setting could hardly have been more tranquil, but the tidings were not. Scientist and writer, we were sitting in a cottage set in the quiet landscape of southern England. From the window of my study, we could see the rolling hills encompassing a small river, which snaked its way to the sea. For centuries, nature and men have molded this landscape together so harmoniously that the one seems a natural extension of the other. The atmosphere was one of such permanence and peace that the disruptive effect of her words caught me totally unprepared.

"Of course I speculate," she said impatiently, "about my experiments and such. Will they work? Have I done them the best way?" She hesitated and then added quietly, "And sometimes I speculate about the new forms of life I could create."

So I faced it for the first time. What science fiction writers have been prophesying ever since Mary Shelley created

Frankenstein, who brought life to a corpse, finally has happened. After two and a half thousand years of endeavor, during which men have tried to understand both the world and their conception of themselves within it, one last cornerstone of the comforting edifice has shifted. Copernicus, Darwin, Lorenz, and Tinbergen transformed man's conception of himself within the universe. Newton and Einstein gave us the knowledge that enabled us to ride to the moon and release the energy frozen in the atom. Now the inheritors of our scientific tradition are ready to become God, the creator, by making forms of life which have never before occurred in the world, organisms which will have the most fundamental property of all living things: they will be self-perpetuating. Once these Brave New Bugs are made, we shall have to live with them.

Already great controversies are swirling around the youthful heirs of Delbruck, Schrödinger, Crick and Watson, the scientific pioneers of molecular biology. From the epicenter of a new technology the shock waves are spreading as public meetings, private arguments, congressional hearings and newspaper headlines all contribute to the greatest debate in the scientific community since the atom bomb. The issue is recombinant DNA.

What is it exactly? Very simply, at its most basic level, it is the new technology that enables a scientist to take DNA from one organism and splice it onto DNA from another to create something absolutely new: new living molecules, new genes and therefore new life. The science is very much in its infant stages—this must be emphasized. More is unknown than known; the repertoire of techniques are still incomplete, the outcome of many of the experiments undecided. But the first beginnings have been made.

It has taken only twenty years to achieve what may well be the last of all scientific revolutions, an event that, in impact, is as momentous as the splitting of the atom. The classic paper in which Francis Crick and James Watson elucidated the structure of DNA appeared in *Nature* in 1956. In the years that followed, our knowledge of the gene and its coding properties grew by leaps and bounds, and the challenge of its mastery acted sometimes like a beacon, sometimes like a goad, impelling scientists forward. Thus it is not surprising that some scientists and technologists now project beyond these Brave New Bugs to Brave New Man, for the knowledge that will enable us to *create* life is the very same that will permit us to *alter* the old. And, so the scenario reads, when we can alter living material at will, Man will be the one animal able to direct his evolution as he chooses. Whereas other species must rely on random mutation and blind chance for their evolution, the human species, by manipulating the first and deliberately eliminating the second, can direct its own destiny. Pick the goal, specify the "ideal" *Homo sapiens* and science will take us there. Such people are not only ready to become as God, they are impatient as well, seeing the next steps in research and application as an inheritance that is theirs by right.

Some of the techniques are to hand; many of the first steps have been taken. We can already fertilize a human embryo in a test tube and, it is claimed, reimplant it successfully in its mother's womb. We can already construct viruses that will cause human tumors by covering a "naked" virus with the coat of another, something neither virus did before. We have already artificially synthesized a gene and set it to work in a bacterium. We have switched on a chemical process in a living cell by inserting an artificial genetic message. We have made a "something"—what *can* one call it—by fusing the cell of a human with that of a tobacco plant. We have made hybrid cells between a yeast cell and the blood cell of a chick; between the cells of a man and a mouse; between the cells of a man and a monkey. We haven't yet grown the adult "something," but this may only be a matter of time and technique. We can already clone: whole plants and animals can be made to grow from single adult cells. We can take the nucleus from the adult cell of a frog, slip it into the cytoplasm of a frog's egg from which the nucleus has been extracted, stimulate it to develop and regain an adult frog in all respects like the one with which we began. The first steps have been taken to do this with a mammal, a rabbit. We have changed the sex of a monkey in its mother's womb. We can choose the sex of our own children. We have taken a five-day old embryo from the womb of one baboon and implanted it into the womb of another, who subsequently gave birth. Son and *both* mothers are doing well. We are seriously contemplating creating hybrids—those "somethings" again—between chimpanzees and gorillas; between gorillas and man. And by "we," I mean *we*: Dr. Geoffrey Bourne, of the Yerkes Primate Center in Georgia, received two letters from the Darwin Museum of Moscow urging him to try exactly that experiment. In Japan, too, a woman is to mate with a primate in another endeavor to raise this hybrid. We are on our way to becoming technical masters of the living process. Nothing will be immutable now, and man himself will be the most potent agent of change. The prospect is exciting, intriguing, irresistible.

But to some the prospect is not at all pleasing. For them the prospects are appalling and dehumanizing and must be resisted. They are like the temptations the Devil offered Jesus, when he took Him up to a high mountain and showed Him the kingdoms of this world, all His to rule. The power within our grasp is an illusion of the most dangerous kind. The notion that we can modify Nature at will brings no satisfaction to those who contemplate the unholy mess—as they see it—that has resulted from man's attempts to modify natural processes so far. Yet others are horrified by some scientists' ambition to manipulate man himself, seeing this as just one more example of professional arrogance and personal ambition. It is hubris of the most dangerous kind, they say, to assume that man needs improving, and claim both to know what these improvements should be, and how to produce them without any dangerous side effects.

In a general, sometimes unarticulated way, more and more people are forcibly reacting against the specter of technological totalitarianism: the haunting juggernaut of a detached, mindless technology based on a vast commercial bureaucratic infrastructure, one that in the name of so-called progress supplants those qualities of humanity and compassion which are as much a feature of man as the logic and rationality on which the technology has depended. Feelings of frustration and anger have surfaced as personal autonomy and identity have been eroded in a variety of ways, and for a variety of reasons, by the wholesale application of what one scientist, Dr. Erwin Chargaff, has called the Devil's Doctrine: *"What can be done, must be done."* The examples are numer-

ous: a biomedical technology that, by allowing technical criteria to supplant human or moral considerations, has pro-
longed human life beyond the point at which there can be
said to be a shred of human quality; a philosophy of research
that, in the name of objective scientific knowledge, allowed
black syphilitic patients in the Deep South of America to go
untreated for twenty-five years; a business technology that
in the interests of profit alone has led to a flagrant ignoring
of the most elementary health precautions—the community
of Hopewell, Virginia, which has suffered irrevocable damage from the chemical Kepone, is but one example; another
is Seveso, in Northern Italy, where in 1976 a cloud of dangerous gas from the Icmesa plant did untold damage. It will
now, they say, be remembered along with Hiroshima. Some
argue that the price for material comforts or even medical
progress has risen far too high. The pressures of the technological imperative for objective scientific knowledge were
beautifully encapsulated in the title of a recent BBC television program on medical research, "Am I Doing This for You,
Doctor, or Are You Doing This for Me?" The bland assumption that all knowledge must have good end points is finally
being challenged, especially as it is becoming painfully obvious that we neither know nor understand quite as much as
we have hitherto been led to believe.

Fairly or unfairly, a lot of blame has brushed off on the
practice of science, adding its weight to a strongly antiscientific reaction already under way in society. The burden of
proof has shifted, and those young scientists who will be
creating new forms of life now find themselves in an ambience totally different from that of their fortunate, distinguished predecessors, who had isolation, status, privilege
and cash. My friends are no longer able to take public good
will for granted, let alone the open money coffers into which
the scientists of former years were invited to dip. No longer
can they dismiss questions about the uses of science by re-
treating to the ivory tower or to the "neutrality" of scientific
knowledge. The paradox is an impelling one: the successes of
biomedical research have brought scientists closer and
closer to issues about the autonomy and rights of human
beings, and those issues now challenge the intrinsic value of
the scientists' work. The very success of a molecular biology

that now enables scientists to create new forms of life has inevitably provoked society into challenging their right to do so. Even the choice of a single research area may carry a value judgment. Those who chose to work on the fertilization and growth of human embryos in test tubes or on aging take a value-laden decision fraught with ethical and societal questions which scientists who choose to study the passage of water through leaves do not have to face. Questions such as: What kind of society would we produce if we extended our life-span for fifty more years? How could we accommodate the population in this already crowded world? How could we provide a life with meaning and quality for all? To do such research is to *know* that one is bringing this situation nearer. The social environment, too, is vastly different from earlier years. In 1967, would any biologists risk saying publicly about recombinant DNA what one physicist on the Manhattan Project is reported to have said about the atomic bomb: "Don't bother me with your conscientious scruples. After all, the thing's glorious physics"?

The current of contemporary biological research is meeting head-on the rising tide of society's involvement with science. The turbulence generated threatens not only to disturb the choppy estuarine waters where scientists have traditionally met society, but even the placid backwaters of the profession itself where, up to now, the only real disturbance came over such issues as to whether or not a person's experiments really did bear out his theories. The new pressing questions are producing angry schisms within the profession itself, for scientists, just as society, hold a wide variety of opinions on such issues as: Do scientists have a divine right to the pursuit of truth? Should the regulation of science be a matter solely for the profession? Just how much should the public be told about, or involved with, research? Do we now need measures other than objective knowledge in judging the worth of an enterprise? Does anyone have a God-given right to search for truth wherever it may lead? Are there some things that are better not known? Whose is the responsibility to guide and control the application of science? If certain experiments are forbidden, what sanctions can or should be brought to bear against those who flout the regulations? What degree of political control over science is necessary or desirable? If such control is exercised, doesn't it lead us directly back to the days of the Inquisition? How can we make the scientific profession more aware of the social, ethical and moral dilemmas—indeed of the very outcome—of their work? How can we make the public more understanding of the enterprise, more willing to be responsibly involved, instead of antagonistic, frightened or envious?

These questions are not new, but it is recombinant DNA and those Brave New Bugs that have forced them out into the open. Had the recent work in molecular biology been as innocent as the problem of the number of lichens accommodated on a church spire, or as abstract as the number of angels upon the head of a pin, the social questions about the impact of science could rumble on like distant thunder, ignored by most scientists who, happily minding their own business, have rarely troubled to lift their heads from the bench. But these bugs are neither innocent nor abstract. The potential dangers may be enormous. They may be highly pathogenic and cause epidemics, or even pandemics, in humans. They may spread into many new environments and cause great ecological damage. They may cause cancer. No one really knows. The potential benefits of the research could be equally enormous. We could, for instance, use the bugs with bacteria as biological factories, to make large quantities of human hormones like insulin, or to make antibiotics. We could reduce our dependence on artificial fertilizers by inserting into a variety of plants the genes that fix nitrogen from the air. We could cure biological defects. Using the bugs will certainly help us understand the gene and may even help us *solve* the problem of cancer. But, again, no one really knows. The risks and promises alike are uncertain factors. But they are of such a magnitude as to generate unheard-of public reaction.

In May 1976 the regents of the University of Michigan called a special meeting to discuss whether or not the new DNA research should be allowed on campus. Was it too dangerous? Was it amoral? The last time they called such a meeting was in 1960 during Black Action Week, when the university was almost shut down. The city council of Cambridge, Massachusetts, where Harvard University is located, argued at length over a resolution by its mayor, Alfred E.

Vellucci, to ban the research for two years until its safety was determined. After several meetings of passionate and often emotional argument, with scientist set against scientist, and council member against council member, a three-month ban was agreed upon, followed recently by a second three-month ban, then the establishment of a review committee of ordinary citizens to help the council decide. In July 1975 the Pasteur Institute, the most cherished and sacred institute in France, "exploded" in anger—to use the words of the report in *Nature*—during an internal meeting of research workers and technicians, who protested both the decision to set up a special laboratory and the experiments themselves.

To their great credit, however, the scientists realized the potential dangers of this work well in advance and, more importantly, did something to anticipate the problems. When information about the hazards began to emerge, scientists were the first to call for a temporary moratorium on research until an international conference could be convened to discuss the issues and agree on guidelines for safeguards. (These guidelines now exist, established under the auspices of the U.S. National Institutes of Health.) This conference, held at Asilomar, California, in February 1975, revealed that the initial problem hid a whole complex of difficult issues, issues fundamental to the practice of science, which would neither go away nor could be swept under the rug. It is nothing new to the scientists that, upon examination, one question dissolves into a host of others: it happens all the time in the laboratory. In this instance, however, scientific questions obscured a host of moral, legal and ethical issues, and the unprecedented degree of public debate caught many scientists by surprise. Their present reaction to the debate is strong and must be understood in terms of a group that feels threatened, that sees many long-cherished assumptions about to be swept away on the wave of what they feel to be an irrational antiscience movement.

In the original legend of Pandora's box, all manner of evils and miseries flew out over the earth after the lid was opened. As Hesiod wrote, Hope alone remained inside, for the lid was shut down before she escaped. Scientists are concerned that the ancient legend will apply again. They have opened their

box, just a little. Society has taken a peek inside and threatened to take the box away, at least until the "manner of the
evil" inside can be determined. Scientists counter this by
saying that the extent of the evil cannot be determined unless the box is opened still further. Society is justified in
demanding that the questions be laid open to public scrutiny,
but some of these are so highly contentious and emotional,
that, as Dr. Sydney Brenner of Cambridge University reminded me, they take on the quality of the arguments in the
United States over gun control: "It is only dangerous in the
wrong hands; self-defense (i.e., the right to pursue the truth)
is a basic right."

The scientists feel that they are standing on the threshold
of a new age. They often feel this, but most especially when
they have achieved the kind of breakthrough represented by
recombinant DNA. But they are also unhappily aware that
something has gone sadly awry with their social contract
with society. The fruits of knowledge are acquiring a bitter
aftertaste and many people are rejecting the methods and
attitudes of science, turning back to humanistic modes of
thought, even to cults of irrationality. The aphorism is now
in danger of being inverted: Art comforts. It is science that
disturbs.

The following pages will examine the legal, moral, societal
and ethical issues of the new science of recombinant DNA.
But first we must understand just what the science is . . . and
what it can do both for good and ill.

2
Creating
New Molecules

*I have read heaps of agricultural and horticultural books,
and have never ceased collecting facts. At last, gleams of
light have come, and I am also convinced (quite contrary
to the opinion I started with) that species are not (it is like
confessing to a murder) immutable.*

—Charles Darwin
Letter to Hooker, 1844

They say that any schoolchild could do it. In that
case, a forty-nine-year-old writer possibly could.
So, with the help of three friends (average age thirty-two), I
learned how to create a new form of living molecule. My
tutor was Peter Carlson, who holds the Chair in Plant
Biology at Michigan State University. Drs. Hugh Robertson
and Gerry Vovis, both of Rockefeller University, supplied
some extra touches, important clues and fine details.

In a pre-"takeoff" briefing, Peter warned me that I needed
two things: time, for the whole procedure could take up to
three weeks; and nonchalance, for the routine is thoroughly
unromantic. I could expect neither sudden exciting action
nor reflective moments of euphoria. There would be no
bursts of light nor clangs of sound, no bubbling cauldrons
nor whizzing instruments. The cries would be those of
"Damn" rather than "Eureka."

"It is not like watching a sunset," he said firmly. "It is more like digging a trench!" To which Hugh Robertson added, "And it's not even like that!"

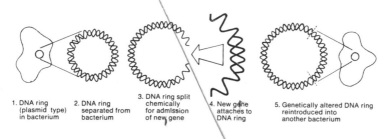

| 1. DNA ring (plasmid type) in bacterium | 2. DNA ring separated from bacterium | 3. DNA ring split chemically for admission of new gene | 4. New gene attaches to DNA ring | 5. Genetically altered DNA ring reintroduced into another bacterium |

The idea is to incorporate a portion of the DNA of a mammal, which is large, into the DNA of a plasmid, which is small. When that is done, the recombined plasmid is allowed to infect a bacterium, *E. coli*, that inhabits the human gut, just as a virus infects a cell, for plasmids are the viruses of bacteria. The bacterium is then cultured in a most favorable environment, whereupon it divides rapidly. Every thirty minutes, one bacterium will have produced two, so at the end of forty-eight hours there are quite a lot of bacteria in your soup. Every time the bacterium divides, so does the plasmid, and so, too, does the mammalian DNA incorporated with it. Thus the bacterium has become a factory that goes on producing recombined DNA until you wish it to stop. That is easily arranged: you merely cease to coddle it.

"What particular recombination would you like to try to make?" asked Peter.

I didn't hesitate a moment. I had heard that one of the likely benefits of this new technology would be that human proteins could be made in quantity, which is vitally needed in medicine. "I'd like to attach the human gene for insulin to a plasmid, and put *that* into a bacterium," I replied.

Peter laughed and shook his head. "I can only take you part of the way," he said. "We can't do that experiment except under very strict containment facilities, to ensure safety, and we don't have them. And even if we did, you've chosen one of the most difficult genes to get. We have, as yet, no way of pulling out that gene, and that one alone, from the mass of human DNA. We might hit it just right, but we might not.

So we'll go part of the way . . . and finish the experiment in
theory, in our chairs!"

To begin with, one needs a plasmid. To get one, you need
a bacterium that already has the particular kind you want.
To get *that*, you need a friend who has made one. Peter
Carlson most frequently uses those made by Professor Stanley Cohen of Stanford University, a pioneer in the recombinant DNA field, so he writes and asks for PSC 101: Plasmid
Stanley Cohen 101.

When botanists discovered a new plant in the courtly days
of science, they would name it after a colleague whom they
wished to honor. Times have changed, and that change began with Francis Crick, the co-discoverer of DNA, who characterized a variety of phages and labeled them with his
initials: FHC.0, FHC.1, and so on. Now, in addition to PSC 101
there are plasmids Charlie Thomas (CT) and plasmids Phil
Leder (PL). By now there are many well-known varieties of
plasmid, with features and properties that are thoroughly
described. They are kept in deep freezes in the laboratory,
and it is as easy to order one up as it is to order a packet of
dried yogurt powder.

On day one, therefore, the mailman delivers a small vial
of freeze-dried *E. coli* of PSC 101. It contains small crystals,
like an instant coffee that has lost its color through old age,
and, just as with instant coffee, we toss the crystals into a
fluid—in this case a juicy nutrient broth containing all sorts
of "goodies": sucrose, glucose, vitamins and mineral salts.
The bacteria love it and thrive. After two days stewing in the
incubator at 37 degrees C, the flask contains a cloudy fluid of
pale beige color, a sure sign of a thick colony of bacteria. It
has a very characteristic smell, too; all colonies of bacteria do,
and an experienced bacteriologist can identify them by their
smell alone. Chanel is never going to bottle any of them up,
however. *E. coli* smells fusty, like the musty gravestones in
a damp neglected church of Northern Italy. *Pseudomonas,*
another bacterium that lives in the intestine, smells like
peaches—rotten ones.

The next step is to get our plasmid out of the bacterium,
and since we never see the plasmid from start to finish, there
has to be a technique to enable us to keep hold of it. It is a

circular object, about one hundredth of a millimeter in diameter—so keeping track of it in the middle of a mixture is not easy. In constructing it, however, Dr. Stanley Cohen has taken care of that problem for us. He has attached a marker to the plasmid in the form of a piece of DNA which ensures that it is resistant to the antibiotic tetracycline. The same drug resistance will be conferred on any bacterium that the plasmid infects. So if we do the final culturing in a medium containing tetracycline, any bacteria that do not have that one resistant plasmid will be destroyed.

By day three we are ready to separate the DNA of the plasmid from the DNA of the *E. coli* in which it is living. By now we have a large supply of both. Bacteria, unlike the cells of higher organisms, have no nucleus, so their DNA is loose in the cytoplasm. It forms a circle, one continuous thread of DNA. The plasmid forms a circle too, but a much smaller one, about one-tenth the size, and quite independent. The smallest plasmid we know contains about twenty genes, containing, in turn, approximately one thousand of the base pairs that are the building blocks of DNA. Plasmid SC 101 contains about nine thousand base pairs, and when we cut its DNA circle and stretch it out, as we will in the experiment, the threads occupy about three thousandths of a millimeter.

To extract the DNA, we must burst open the bacteria which house it. The colony is first reduced from a nasty-smelling fluid to an innocuous pellet by spinning it in a centrifuge with a force of several thousand times that of gravity for ten minutes. The pellet is then either compressed viciously or washed drastically. In the first case, it is delivered over to the mercies of the French press, a name which aptly reflects the kind of medieval torture the bacteria are about to undergo. They are blasted, one by one, through a space as narrow as the finest hypodermic needle, with a force of several thousand pounds per square inch. The bacteria pop open as they come through, and their innards are splayed all over the flask. It is a crude method, and since the sheer force can break the DNA, it is only used when it is not necessary to keep the DNA intact.

We choose a more delicate method of destruction and use chemicals that digest away the exterior of the cells. The pellet is placed in a test tube with various other fluids: enzymes that chew away at the proteins, carbohydrates and fats; a detergent that strips the membranes off the bacteria; and a buffer solution that holds the fluid at the right acidity-alkalinity balance, so that all the biological reactions are stopped. The temperature is again 37 degrees C.

It is now three days since the mailman called, and the test tube contains completely disorganized bacteria. Now is the time to extract the DNA. By adding very cold ethanol to the mixture, the DNA can be concentrated in layers. After the mixture is centrifuged again, everything except the DNA floats to the top and is poured away. Then, in a wonderful moment, we "spool out" the DNA. The tube is kept cold and a glass rod is spun around by hand. Slowly long threads adhere to the rod and when you pull it out, it looks like a miniature stick of candy floss. "But if it's pink," said Peter, "we're in trouble." The white threads of pure DNA can be seen for the first time, and at last we can begin to separate the plasmid DNA from the bacterial DNA. Since the bacterial DNA molecule is at least twenty times bigger than the plasmid DNA, the two can be separated by their size or density.

This stage provides one of the few real aesthetic moments in the whole three weeks. The key ingredients are flat nine-inch-by-twelve-inch squares of thin gelatin called acrylamide; one might be making eggs in aspic. In fact the whole procedure has a strong cookbook air about it. "Take so much of this; add so much of that; blend . . . or spin." The instructions are this: put a small drop of the mixed DNA in the small cavities at the top of the gel. Pass a current through the gel to pull the molecules down. The bigger and heavier the molecules, the slower they will travel. Leave for about two hours. When you are done, the heavier, bacterial DNA will be at the top, the lighter broken bits at the bottom, and some two-thirds of the way down will be the plasmid DNA from PSC 101. You still can't see them though, so stain everything a faint orange, with ethidium bromide, and wait. In a dark room, eyes shielded from the ultraviolet light that must be beamed onto the gel, we see the evidence slowly emerge. Out of the background of the gel lines of DNA begin to appear. They look like a series of dark metal staples that have been clamped into the matrix of the gel, and some are much thicker than others. The sight is wonderful; the moment a powerful one.

Scientists have done this procedure so often that they know which line represents the plasmid DNA; there is no need for hours at the electron microscope looking for the telltale circles. One can go straight to the next step.

During that week, we would also have been preparing the mammalian DNA that we wished to attach. As I had been warned, this was a considerably less exact proposition. We would begin with about ten grams of human cells ordered from a drug company, and grew them in tissue culture. The DNA was extracted in precisely the same way as the bacteria's, with enzymes and detergent. However, now the problems really start, for in *that* test tube not only is there enough DNA to code for fully a million genes, and the one gene we want—for insulin—is about one to ten thousand times more complicated than the bacteria, let alone the plasmid we had been working with. We don't know yet how to pull that gene neatly off from the rest. The method is rough and aptly called the "shotgun" technique. This means chopping up the human DNA quite indiscriminately, and hoping that some of the bits will both contain the gene we want and be captured by a plasmid. It is not as big a gamble as rolling dice, for something *will* be captured, but we still don't know what. The bemused scientist does the experiment, then says, "What have I got?"

It was now day six, the end of the week. Peter would take the weekend off to be with his family, but at Rockefeller, he insists, others are likely to work straight through! So it is on the Monday morning that the work continues. The two test tubes of DNA come out of the refrigerator, and the next stage, splicing the DNA, begins. Our "scissors" have been around since life began. Bacteria and viruses live by parasitizing other organisms, by insinuating their DNA into their hosts in order to replicate and survive. For this purpose, they have evolved substances that allow them to slice up foreign DNA. Once they have sliced their way in and safely latched on, they use other enzymes to repair the splice, sealing themselves into place. The slicing enzymes are called "restriction enzymes," and they are very specific: each one always makes its cut at the same point sequence in the DNA, and the cut leaves the ends "sticky." As a result, any DNA, whatever its source, bacterial or human, will join up with any other DNA, from whatever source, *providing they have both been cut with the same restriction enzyme.* It is this one marvelous property which makes the science of recombinant DNA possible at all.

We add the same restriction enzyme to both test tubes, and it proceeds to cut up the DNA. The enzyme acts like pinking shears: instead of making a neat slice at right angles to the threads, it cuts diagonally across the two. It is this "staggering" of the cut that makes the ends "sticky," for the bases that it cuts are complementary to each other, and when they find each other again, they join. The preparation is kept buffered in a solution that keeps the molecules tranquil and inactive. The plasmid DNA, once cut, opens up; its circle becomes a straight single thread. Every plasmid has been cut at the same point in the DNA sequence, so the test tube is full of plasmid threads, all alike. By comparison, the human DNA would be a mess. In that test tube are thousands upon thousands of bits of DNA, of all possible lengths, carrying all sorts of genes with all sorts of functions. Yet every one of those bits will join both with another human bit *and* with the plasmid thread, once the buffers are removed from the solutions and the contents are mixed.

Were we permitted to proceed further, day ten would be the day of the conjuring trick; "a piece of pure 'magic,' " Peter says. It would go like this: in each hand I hold a tube of DNA, one containing plasmid genes, the other human, yet looking exactly the same. The trick is to persuade the plasmid to capture an interesting piece of DNA. But, sadly, the plasmids will capture at random. When they have done this, we will insert them into bacteria. Only then can we begin the hunt for the bacterium that has the plasmid that has the insulin human gene inserted into it. It is, as Peter says grandly, "a magnificently difficult task." Julia Child would have to metamorphose into Sherlock Holmes.

The fluids would be mixed together in a low temperature that favors the joining of sticky ends. From the moment we extracted the DNA we have been working, not with living organisms, but with single chemical molecules. Keeping everything at a high temperature has held the living activity at bay, but the moment the temperature is lowered, the quality of life reasserts itself, and the DNA threads begin to pair up again; to do that, they have to find a mate.

The tube is put into a water bath so the pieces of DNA begin to join. We add the DNA "repairing" enzymes that will anneal the pieces tightly, and it is done. We have recombinant DNA. The test tube contains a very mixed bag, however. Some plasmid has stuck to plasmid; some human to human; some plasmid to human; and in this last category, *somewhere,* are the plasmids that *may* have captured that piece of human DNA, of unknown size, that has the region that has the gene for insulin. We will need a messenger, a probe to go in and get it out. There are thousands and thousands of bits of DNA around, so it is like asking someone to go into Times Square on a crowded Saturday night and fetch out of the crowd one person whom he doesn't know, and whom we can't accurately describe. Our messenger will have to scan everyone. But at least there are some clues for Sherlock Holmes, and at least we can get nearer to our plasmid. The next step will be to get the plasmids back into *E. coli* and end up where we started, with freeze-dried plasmid. But now the plasmid will contain a piece of human DNA. This recombined DNA will not form a small plasmid circle, but a whopping great big one. It may contain more than one human gene, but there's nothing we can do about that.

First the plasmids must be made to infect an *E. coli* population known to be sensitive to the drug tetracycline. This time a combination of low temperature and high calcium is sufficient to do the trick, and *E. coli* greedily absorbs all the DNA around. This second *E. coli* population now contains some bacteria with plasmid gene alone and some with human gene alone—and some bacteria with recombined DNA plasmid inside.

These populations will have to be sorted out. About one thousand cubic centimeters of the mixture of *E. coli* will be incubated as before, in a small glass dish containing tetracycline. Those bacteria that don't have the drug-resistant plasmid inside simply won't grow. Those that do form, in forty-eight hours, small, pale-yellow colonies on the medium, which can actually be seen with the naked eye, and picked off on the end of an ordinary toothpick.

Basically this is as far as any scientist can get with certainty in his search for the gene for insulin. We would like the gene to reveal its presence to us, but the only way it could do that would be to show us its biological activity. We would know if the gene for insulin were there if insulin itself were produced. But so far it hasn't been, and we don't know if it ever will be. This is the single greatest obstacle to the exploitation of recombinant DNA techniques. The problem is in getting the genes of higher organisms to become fully expressed in what for them is a totally foreign environment. We are asking them to function normally in a situation where the control mechanisms are abnormal. Many young scientists working in this field realize that they may be spending a lifetime struggling with a problem that has no solution.

In bacteria, things are much easier: gene swapping and gene repair is simple. Moreover, taking a bacterium that is defective in one function and making good that function by using a recombined plasmid to deliver the missing gene is now child's play. A British husband-and-wife team, Drs. Kenneth and Noreen Murray, acknowledged as masters of the field, can, I conclude, do almost anything they choose with bacteria and bacterial genes. But mammalian genes are a different story. They are so much more complex.

To the outside observer, the whole procedure demands an incredible act of faith to be sustained through a game of elimination and numbers and biochemical acts. Gerry Vovis said: "It's a noble tradition not to be able to see what you are doing." It may be, but it requires a lot of self-confidence! One has to go on and on and on, blindly doing the procedures and manipulations, confident that the end result will be what you want. It works because over the last thirty years biologists have built up a fantastic tradition that combines the best of genetics, biochemistry and microbiology. They have developed a series of finely tuned selector systems to enable them to pick out those rare microbiological individuals they want, or have created: in this case a plasmid with the foreign section of DNA.

Having made themselves technical masters of that little universe far below the level of man's vision—the world of bacteria and viruses—scientists now seek to extend their skills and command to mammalian cells, to problems at least three to four orders of magnitude more complicated than anything they have tackled before. And yet, for all the uncertainties, they are succeeding in this, too. Had I chosen, for instance, to attach a hemoglobin gene from a rabbit instead of an insulin gene from a man, the experiment would have been very precise, nothing shotgun about it. There are blood cells in the adult rabbit that make nothing but hemoglobin, the protein that carries the oxygen in the blood. Though the cells carry the full complement of rabbit DNA, much of it does not function. The gene, with its own control regions, remains intact in a significant portion of the spliced fragments, and can also be hybridized with its own messenger in much of the bacterial population into which it has been inserted. Its presence is thus revealed. Once revealed, the bacterium that have it can be selected out. The technique is no longer "shotgun," for the scientist knows precisely what he wants and whether he has got it.

So what does all this mean?

There is no doubt that any schoolchild could do some of the simplest types of experiments, given access to a laboratory and a guide, a catalogue of the drug companies and some cash. The technique in bacterial systems is routine now for first-year biology undergraduates. Anyone who wishes to have his own do-it-yourself–recombinant DNA laboratory at home could easily equip himself for about fifty thousand dollars. But why should anyone want to? What does this technology mean for us? The answer to this, of course, depends on who you are. Dr. Charles Weissmann, one of the most distinguished scientists in this field, provided one answer when I spoke with him in his office at the Institut für Molekularbiologie of the University of Zurich. I had asked him why, if the benefits and dangers alike were equally hypothetical, was there so much fuss? Why did scientists care so passionately about recombinant DNA?

Gentle, funny, charming and irreverent, he stared at me in amazement. "But the whole scientific world has gone crazy about it," he said. "I know of no other major laboratory that's not thinking about the technique, or doing something with it." He leaned forward on the desk, smiled at his own exuberance, and proceeded to educate me. "You see," he continued, "we thought so much was condemned to be a mystery. In bacteria the study of genes and their expression is now a trivial problem. We understand how they are regulated and the steps by which the protein is made. In the early, golden age of molecular biology we were confident that if we understood *E. coli* we understood it all. That confidence didn't last long, for the higher organisms have many, many more genes, volumes more DNA. To isolate one of these genes cleanly enough for study means isolating one part in a million, even one part in ten million. You'd need a trainload of the stuff and you couldn't live long enough to process it.

"So everyone was screaming, 'Something new is needed!' and this technique is *it*. Since we can chop the DNA up into little pieces we hope to get hold of single genes, eventually. Then since we can multiply them up, once they are in a bacterium, we hope to get as much of any one gene as we need. So we are at last working at a level and a scale that can be easily handled in the laboratory, and we can see our way to understanding how they work."

Dr. Fred Niedhardt, chairman of the Department of Microbiology at the University of Michigan, gives it a more philosophical twist. "For me, understanding how cells are put together and grow isn't far removed from my asking a lifelong question. What is life? Who am I? What is the universe? What is God? What is the nature of what we perceive? I see myself able to comprehend answers at the cellular level, to be able to contribute to the formulation of the questions and getting some of the answers. I'm lucky. I'm enjoying that game, and it is part of the much larger game of man's endeavor to learn more about himself. I would be upset if somebody walked into my room and said the work on recombinant molecules was judged to have far greater risks than the potential benefits, and that the opinion of society would be that they would not support it. But I would be really shocked if somebody walked in and said, 'We don't want to learn any more about a cell, and your ultimate goals are what we don't like. We do not feel that man should know that much about biological systems.' Because when you talk to someone whose life is wrapped up in science you find that they are in it because of their curiosity. That is what has motivated them, and it is no different from asking the same question of a writer like yourself or an artist or a religious thinker or any individual who is engaged in creative activity. But just as with these other individuals, society then goes on to reap the benefit of what is created. The individual members of society benefit by having available to them information, understanding, in some cases amusement. And so, I would include *all* this as part of mankind's growth and knowledge about himself."

For many of the scientists the goal is pure understanding. They see themselves as heirs to a gloriously civilizing tradition in man's history. But ordinary people, who may be forced to make decisions about this research, are likely to consider things more prosaically. They will want to know its practical value and likely outcome. At present the benefits may be hypothetical, but they do represent a real prospect for good to human society. When I asked Fred Niedhardt to sum up the likely practical results given the recombinant DNA techniques, he said he would need paper, pencil and a few hours. If he got down to it, he felt he could probably come up with a list of some two thousand items. It would be a fascinating exercise of imagination, trying to anticipate what could be produced if we programmed cells to make the things *we* wanted them to make, rather than those evolution had selected them to make.

In principle there is nothing new about this. Man has always used cells to make useful things. Yeast cells have been used to produce a whole range of delightful products from beer to bread. Fungi are used to grow antibiotics; multicellular organisms can supply food products or even textiles. Animal husbandry, agriculture and the microbial fermentation industry has, throughout recorded history, exerted gentle but directed pressures on biological systems, persuading them to evolve in the direction of being more productive. The milk cows of today probably wouldn't survive very long in the wild because they have been selected for many generations for optimum conversion of hay and grass alone into milk. All this has been a fairly gentle selection, however. Now, with recombinant DNA, genetic tools exist to improve our ability to push evolution along greatly, particularly microbes, and our imagination can now extend beyond the products made by single cells to those of whole organisms and even their genes.

One example will already be familiar from Dr. Weissmann's statement and my experiment. In trying to attach the human gene for insulin, Peter and I hoped to use bacteria as a factory for producing large quantities of mammalian genes and insulin. In theory this could be done for any gene, and any substance, having a major effect on biomedical research. For instance, we would like to know a great deal about the substances and function of particular enzymes, those substances that control many bodily reactions, but a great deal of effort at present is going into the simple problem of getting enough of *any* one enzyme in a sufficiently pure state to do any biochemical studies at all. Most enzymes are present in cells in minute quantities on the order of one-tenth, or even one-hundredth, of 1 percent of the total protein available. If by recombinant DNA technology scientists could introduce a gene that makes an enzyme into a cell, and then make hundreds of copies of the gene and cause that gene to be expressed, the chosen protein could come out of the cell as one in every ten molecules, instead of one in every ten thousand. In addition, making proteins in this quantity would make many studies much easier—i.e., X-ray crystallography, which tells us about the structure of proteins—and give scientists a whole new spectrum of cellular products to look at.

The ability to produce human gene products in all their variety would have wide applicability in fighting disease. Many genetic-deficiency diseases can be mitigated simply by supplying the missing enzyme. Diabetes is the classic case. The technology for using insulin is by now well established, but it is pork, goat or bovine insulin that is made and used, and insulin from these other animals is just sufficiently different from that of humans to cause severe immunological problems at times. The human patient may make antibodies against the animal hormone and thus suppress its action. But if scientists could get the genes for insulin into a "DNA bacterial factory," doctors could use human insulin for their patients.

Yet another important substance needed in quantity is interferon, a gene product widely regarded as crucial to the body's resistance to viruses and especially useful where antibiotics have no effect. It may, indeed, be important in our resistance to cancer; the Memorial Sloan-Kettering Cancer Center in New York has undertaken an enormous program in interferon research, drawing in scientists from a large number of countries. Once again, however, interferon is made in minute quantities in the body, and each person needs all he or she can produce. Making interferon through a bacterial factory would greatly reduce the vast chemical technology that will be needed to synthesize the substance in sufficient quantities to make it a useful therapeutic tool.

Recombinant DNA techniques might also be useful in the treatment of the genetic disease hemophilia. Doctors now have a perfectly adequate replacement for the factor which is missing in the blood of hemophiliacs, but the only source is human blood. It costs about twenty to thirty thousand dollars per year to supply enough of the factor to treat one patient. Again, if a bacterial factory could be established, scientists estimate they might bring down the cost of this treatment to something like ten to twenty dollars per year per patient, and still produce all the factor that is required. Immunoglobulins, the substances that confer immunity, could be produced in quantity by the same technique.

Aside from the possibilities of interferon, cancer research could benefit in other ways from recombinant DNA technology. I discussed these with Dr. Renato Dulbecco of the Imperial Cancer Research Fund, London, a recent recipient of the Nobel Prize for his work on one group of viruses that may cause human tumors. He emphasized that these techniques would not lead directly to a cure for cancer, but would substantially further cancer research in a number of indirect ways. In particular, these techniques will simplify the study of cancer viruses. For instance, we could "amplify" those DNA segments that are of research interest, thus helping us identify the genes that are crucial in the transformation of cells from a normal to a malignant condition. Scientists have a fairly good idea which protein in the SV40 virus changes a cell from normal to cancerous, for example, but at the moment it is literally impossible for them to get enough of the protein to do the experiments they need to find out how the change occurs. Putting the crucial gene—and *it* alone—

into a plasmid means once again that a bacterium could be forced to produce it in quantity.

The possibilities are almost endless, and the next two chapters will look at more of them. However, there is another face to the research as well, the one that has provoked so much professional and public concern: its inherent dangers. For the same technology that could create vats of beneficial molecules could possibly also produce those that are destructive, which might irrevocably change the biological balance in our world.

We can best understand these problems by seeing the context in which they first arose. This means going back to the very earliest work, to Stanford University in California and to the laboratory of Dr. Paul Berg, one of the scientific and moral leaders in the field, and a central figure in this whole story. The story was told to me by Dr. David Jackson of the University of Michigan, who was in from the very beginning. Thirty-three years old, serious, articulate, and luckily for me a most skilled expositor, David Jackson has a reputation both for brilliance in science and a strong sense of responsibility toward society. This has, according to Fred Niedhardt of his department, "trapped him into being a spokesman for many of us, and for many other people, in recombinant DNA." In the last two years, as the issues have been debated widely, as the demands of public understanding and accountability have intensified, he has found his time for research increasingly eroded. Many young scientists in the field are very resentful about the extra role they are forced to play. David Jackson, in contrast, is not, for he sees it as not only necessary but positively desirable that the public should be given the clearest understanding of the science and the problems it raises by the practitioners themselves.

After graduating from Harvard, David Jackson and his wife went to Stanford to do their doctorates with Charles Yanofsky, whom Jackson regards as one of the best molecular biologists in the world. He got a little ahead of his wife, earning his Ph.D. in 1969, so while she finished her thesis, he went across the road to the Biochemistry Department of the Stanford University Medical School for postdoctoral work with Paul Berg. This move was crucial, because it was the work done there that ultimately led to the techniques of recombinant DNA, then to the recognition of the inherent dangers, and finally, to the scientific moratorium and the Asilomar Conference.

It was a joint experiment by Paul Berg, David Jackson and Bob Symonds, then on sabbatical leave from Adelaide University, that first demonstrated that recombinant techniques were possible at all. During the 1960's it was observed that certain phages conducted natural genetic engineering, routinely picking up DNA from the bacteria they infected, and becoming recombinant hybrids in the process: "Going into the bacteria," someone observed to me, "as virgins, but coming out as bastards." During these years several scientists, most notably Alan Campbell, Gail Kaiser and David Hogness of Stanford, managed to induce the process artificially, and thus provided a marvelous mechanism for studying the detailed genetics of bacteria.

No such recombinant techniques existed, however, for the cells of higher organisms, where by comparison the study of genetics was extremely gross. Whole chromosomes could be exchanged, but there was no technique for studying the genetics of higher organisms at any finer level, let alone a molecular one. So Paul Berg and his colleagues took a small set of bacterial genes and recombined them with the DNA of a very important virus, SV40 virus, which regularly infects mammalian cells, by incorporating *its* DNA with that of its host. Relying on this property, the scientists argued, they would have a technique to insert small pieces of foreign genetic information into mammalian cells in culture in a stable and inheritable form.

Suddenly, however, the scientific implications became very serious. SV40 virus is a proven tumor-causing virus in mammals. Berg and Jackson had talked for some time about their methods, so their plans were more or less well known in the scientific community, but as a successful outcome became more and more likely, people began to think about the consequences and whether it was, indeed, an experiment that should be done at all. Here they were putting genes from a tumor-causing virus into *E. coli,* the intestinal bacteria that grows in human beings. *E. coli* replicates in enor-

mous numbers; the possibility existed of a real public risk if
their population of *E. coli,* now infected with SV40 virus,
escaped from the laboratory and spread.

Some scientists telephoned the group and tried to persuade
them not to try the experiments: they were far too dangerous, they said. Others called it a tempest in a teapot. The issue
was hypothetical since no one knew—and no one knows
even now—what the real risks were, *if indeed they existed
at all.* Many felt there was no reason to think that SV40
would be dangerous in human *E. coli;* no reason to suppose
that the virus would leave the bacterium and penetrate
through the wall of the human intestine to cause cancerous
changes in the cells. A lot of people had already been exposed
to SV40 virus in earlier contaminated polio shots, and no
epidemiological studies have *ever* suggested a greater incidence of cancer in people who had had those shots. In addition, many scientists had been working with SV40 for a long
time without having taken any precautions whatever. Nevertheless, the question remained: what if the new molecule
did escape? The dangers were sufficiently scary for Berg and
others to take the steps that, ultimately, led to the Asilomar
Conference. (The full story is told in Chapter 5).

The possibility that cancer-causing particles can be transferred into humans is perhaps the most dramatic danger, but
also dramatic is the possibility that a brand-new chimeric
molecule might itself be tumor-causing. Combining genes
from organisms that have never been joined before might
produce an aberrant result—an organism with unknown
properties which could be suddenly quite virulent. Besides
cancer, it is possible that two pieces of spliced DNA which
are quite harmless in their parent organisms could together
form a dangerous pathogen, giving rise to devastating epidemics or pandemics, world-wide outbreaks of disease.

In addition, there are genes which confer on bacteria a
resistance to drugs, in particular to antibiotics. Such genes
could also be fused with the normal genes of a bacterium,
and multiply as the bacterium does. A whole population of
drug-resistant bacteria could be accidentally released and,
again, incorporated into human beings. The problem of
drug-resistant bacteria in therapy is already horrendous
enough without the addition of new populations and strains.

A new molecule might also confer new properties upon the bacteria into which it is injected, properties that would then allow that population to colonize a far wider range of environment than is normal, so that, for instance, *E. coli* now living in the gut might be able to live in other parts of the body—perhaps the cardiac muscle of the heart—or its range might be extended to other animals and plants.

How real are these dangers? Some scientists say that K12, the strain of *E. coli* used in the laboratories, is too weak and refined to cause anyone any trouble. K12 appeared on the scientific scene in 1922, taken from the feces of a diphtheria patient in Stanford, California, appropriately enough. But the immune system of the original patient would no longer recognize *E. coli* K12 because its surface coat has gone through a lot of changes which make it defective. In ecological competition outside the laboratory, its chances of success would be remote. It might settle down in the human gut for a small, transient period, but no more. On the other hand, though it may be reasonable to be this sanguine, many scientists would prefer not to take the risk, especially since so many of their colleagues have been shown to be downright sloppy and careless in their laboratory procedures. If the dangerous bugs could be contained in the laboratory, all well and good. But few scientists have been trained in the safety methods familiar to all medical bacteriologists. There are some horror stories already: one scientist, for instance, has given himself a shocking dose of radioactivity by pipetting radioactive fluid by mouth, which is forbidden, instead of by hand.

The dangers of these techniques are not confined solely to the possibility of individual physiological harm. As a silent weapon in biological warfare, pathogenic, newly created bugs may be cheaper to use and easier to exploit. No one knows. As we shall see in the next two chapters, the whole question of gene therapy and genetic engineering in plants and animals raises other forms of questions, too, and the undoubted scientific benefits that the techniques will bring have to be measured against the moral and ethical problems that will follow.

This scientific century has seen an atom bomb and napalm; new strains of rice; a moon walk and a Concorde; vaccines and the elimination of smallpox. Are the consequences of recombinant DNA research on a par with these earlier momentous events? Certainly the prospect of creating new species has provoked in a few scientists, and in others outside the profession, the same kind of agonized scruples that afflicted Charles Darwin when he contemplated the consequences of his theory: "It is like confessing to a murder." The problems we now face are similar to Darwin's in that we are called on once again to balance scientific issues against ethical, moral, even spiritual, issues. Yet in another sense we face a more awesome situation, one which Darwin could never have imagined. Species, he realized, can change. Species, we realize, not only can be changed, but can be created, and man is the Creator. The consequences will be irrevocable. So do we do it, or do we not?

The question may be already asked too late, for some scientists have begun the first stages in the creation of totally new species, let alone molecules.

3
Creating New Species

"When he has learnt that bottinney means a knowledge of plants, he goes and knows 'em. That's our system, Nickleby: what do you think of it?"

—Charles Dickens
Nicholas Nickleby

There has been a very noticeable pecking order in the sciences in the last hundred years. Physics established itself somewhere near the top: a place physicists have always felt was theirs by right anyway, since they argued it was only a matter of time before all other sciences were subsumed under their rubric and everything in the world explained in terms of the forces between fundamental particles. From that stance, the biological sciences had been correspondingly well down the order, the study of plants occupying a place even more humble than that of the study of animals. As for natural history, those observational studies which, as it turned out, were to provide the single most important basis for contemporary ideas about ecology and animal behavior—it was once scornfully dismissed as the occupation of the amateur: an activity or hobby for those who were either not qualified to be "proper" scientists or did not take science seriously.

In many respects, the shoe is now on the other foot. Professor Evelyn Hutchinson, one of the most distinguished ecologists of the twentieth century, who retired from Yale University a few years ago just when the ecological movement was surging towards its high tide, remarked wryly how delightful it was to learn that one had been doing a fashionable and important science all these years. Ecological studies find favor not only because they produce empirical results, but also because they have forced us to take a more perceptive and sensitive view of man and his relationship with the world in which he lives. They are practical too: one foot is always firmly planted on the earth, quite literally, and botanical and agricultural studies have benefited the Third World countries for a long time. Even the ultimate accolade, the Nobel Prize, has been awarded in recent years to Norman E. Borlaug, in 1970, for his work on the Green Revolution, and to Konrad Lorenz, Karl von Frisch and Nikolaus Tinbergen, in 1973, for their work on ethology. This has admittedly taken a semantic sleight of hand, since the Nobel Prizes are technically restricted to physics and chemistry, or physiology and medicine (Borlaug was actually awarded the Nobel Peace Prize). However, it is a sign of the growing recognition of the importance of these observational sciences that the prize committee belatedly performed those particular conjuring tricks.

Thus it might be paradoxical, but it would certainly be appropriate, if we found that the first beneficial pay-offs from recombinant DNA research and genetic engineering were to come in the fields of agriculture and botany. Botanists have been doing clever things with genetic manipulation for a long time. For generations, breeders have selected plants and animals with identifiable and desirable traits and bred them to produce new characteristics and, indeed, new species. Like all technologies, this has had its good and bad effects: the hardy strains of plants and the high-yield strains of wheat are a worthy monument to these endeavors. The lapdogs of Manhattan and London, bred for more dainty characteristics, are not. Now, however, botanists and most especially plant geneticists are on the verge of even more remarkable breakthroughs: the ability to create new species entirely by genetic manipulation, as opposed to selection; to confer new properties on old plant species; to extend the range of environments in which plant species can live; indeed, to do all of these things to one plant.

For many years plant geneticists were in the doldrums, that condition delightfully defined by the *Oxford English Dictionary* as "non-plussed." There had been a theoretical and practical stalemate, and genetics were possible only on the gross scale. But now, the importation of microbiological techniques into the science as an offshoot of the earlier studies on phage and bacteria has given plant geneticists a tremendous tool. Far from being in the doldrums, this branch of botany is sailing fast under the influence of a very steady wind.

This has happened just in the nick of time, for we are facing some extremely difficult problems in agriculture, and unless man alters his habits, even his habitats, drastically— which is unlikely—science will have to come up with new solutions. One of the things we need badly is a technique for the rapid development of more genetic variability in plants: that quality of being genetically different which is reflected in such obvious characteristics as tallness, disease resistance and so forth. We need it because the standard arsenal of genetic variability hitherto available in nature is dwindling rapidly. Variety is the vigor, the mainspring of life, the basic material on which evolution works. Without genetic variability there can be neither strength nor possibility of evolutionary progress.

The crop plants that men use now and have used for thousands of years evolved in Asia Minor, China and South America. In some of these areas the original ancestral wild plants are still to be found. These plants have a tremendously wide range of germ plasms, covering an infinite variety of properties. The very success of our agriculture in designing crops with important *single* traits, however, has meant that over the last five to ten years, farmers all over the world have been planting more and more of the "improved" varieties of crop plants. These are monocultures, forms from which the variability has been bred out. The traditionally genetically diverse crops are being planted less and less, causing our store of natural variability to be rapidly depleted and irrevocably lost. There is another element as well. For obvious reasons people have traditionally built cities in river valleys: the flat fertile plains provide the best habitats and transportation is easy. River valleys are also places where there is an abundance of natural plant life, however, and therefore too, the places of great genetic variability. As a result of agriculture becoming increasingly mechanized and society increasingly urbanized, the range of genetic variability has dwindled almost to vanishing point.

It is desperately urgent to conserve genetic variability for its inherent vigor and as a source of favorable characteristics, and also because it is the only resource to which we can turn when a monoculture is threatened by a new disease. In spite of its positive benefits, the Green Revolution brought with it a number of disadvantages. The disease problem was one such: one new race of rust fungus or one new bacterial disease could wipe out not only a majority of the crop but also a country's agricultural base. When a new disease evolves, where else can we turn to find resistant varieties except to the genetic variability present in wild populations? Until recently, the international collection centers for plants had not kept pace with the need to collect and preserve this germ line. In the same way zoos now find themselves performing the function of keeping and rearing animals which face extinction in the wild, so we shall shortly see "plant zoos" that will perform a similar function for plants. At Fort Collins in Colorado, some 90,000 pans of plant seeds sit in cold storage in the stone and cement building of the National Seed Storage Laboratory. They may prove to be more valuable to the world than all the gold stashed away in the vaults of Fort Knox.

We have become far too dependent on plants with a very narrow genetic base. Ninety percent of the soft white wheat grown in the U.S. Pacific Northwest comes from just two varieties of high-yielding hybrid seeds, seeds which are promoted and sold in other countries as well. So too with other crops: the grain sorghum grown in the American Midwest for animal feed is the same as that grown in parts of Africa. The corn crop in the United States in 1970 was extensively damaged by blight, largely because the hybrid corn had little or no resistance to it, and 20 percent of the crop was lost— the feeding equivalent, it has been calculated, of 32 billion

McDonald's hamburgers! In many respects the situation is more worrying in America than in any other country. Current agricultural species all come from different areas of the world, and none are indigenous to North America: there is no native plant stock, for all cereal crops grown in America came from the East. Even those raised by the American Indians originated in Latin America. Scientists now estimate that two-thirds of the early American oat varieties, 98 percent of the clovers, and 90 percent of the soybean strains have been completely lost.

Scientists would ideally like to have some 10,000 seeds from each variety. The agronomist who opened a can of a rare tomato species and found only twenty-three seeds got a rude shock. There are at least three problems facing agronomists here: that of inadequate storage facilities; that of collections which are, quantitatively, also inadequate; and that of purity—the identification of collections can't always be trusted.

Many scientists believe that merely saving species is only one priority. A second must be to develop techniques which themselves generate variability, and it is here that we see what plant geneticists can do. Professor Peter Carlson is himself one of the brightest people in this field in one of the most active centers for this work, Michigan State University.

It was at Yale University that he learned the techniques of plant tissue culture—appropriately enough, since Yale pioneered tissue culture studies in the 1930's. At that time a zoologist named Ross Harrison developed the basic techniques that led to the wholesale use of tissue culture with animal cells, and thereby transformed the possibilities of biology.

There are, Peter told me, a lot of things one can do with plants that it is quite impossible to do with animals. One of the troubles with animal genetics is that it is impossible to get a viable cell with only one set of chromosomes. Apart from eggs and sperm, all animal cells have, and can function only with, a duplicate set of chromosomes, one set from the father and one from the mother. Characteristics in offspring, such as eye color, are ultimately determined by which of the parental genes for eye color is dominant: brown over blue. Unfortunately, the brown-eye gene is generally the only one whose effects you can see; the blue stays hidden unless the child has a gene for blue eyes from both parents. Since most changes—"mutations"—in genes are not dominant, variants which can be recovered are rarely the ones you want. To use Peter's words, "Genetically, animals form a thoroughly mixed up population."

Plants, however, make things easier, for they are much more plastic with regard to their genes and chromosomes. Many plants can grow where the chromosome set has been not only doubled but quadrupled. Plants will even grow whose cells have only one set of chromosomes: they are called "haploid" plants. Since there is then but a single copy of the gene, with nothing to dominate it, mutations can be easily spotted. In addition, scientists have learned how to regenerate whole plants from just one single cell. It has been done with carrots and with tobacco plants. Here scientists have been able to become really ingenious: they retrieve single cells of tobacco plants, with only one set of chromosomes, and manipulated them genetically to give them new properties, such as resistance to herbicides and disease, or to environmental stresses such as heat and cold. Having manipulated the cells, they then grow whole plants again. They can also take a haploid tobacco plant, culture its cells, treat them genetically, regenerate the whole plant, take one of its cells, double the chromosome number with the drug Colchicine, and then grow another plant. Thus from a single haploid cell, which has been modified in culture to suit their fancy, they can get back a normal adult plant with the standard number of chromosomes, but genetically altered. This is an extraordinary achievement—and quite impossible with animals.

This kind of analysis gives botanists a very valuable tool for refashioning plants. For instance, scientists can now fuse one plant cell with another. Plant cells are surrounded by cellulose walls, little prisons of fiber which allow the contents neither to escape nor to fuse. However, if you strip off the walls, either mechanically or with enzymes, you can fuse together the contents of two plants and produce a hybrid plant cell. In this way, you circumvent all the problems of sexuality in higher plants. The cells of higher plants, like higher animals, will not readily fertilize across species—the pollen won't combine with the eggs to produce viable seed —but by using the cells of the plant, any plant cell can be made to fuse with any other cell, no matter what species. Cells are downright promiscuous. As a result, botanists now fuse cells from a wide variety of plants, stimulate growth and produce a wide variety of hybrids. It is even possible to fuse the cells of the duck with those of the orange for its sauce, or with those of an ostrich—though it is not yet possible to get the hybrid to grow!

Most of Peter Carlson's work, and that of his laboratory at MSU, is to study those conditions that will permit him to regenerate important crop plants from a single cell. The intellectual game has already been played out on tobacco. The real task now is to shift gears and start again, to try to do the same things with wheat, corn, barley, oats, rye, beans and other vegetables—once again, aiming to increase the range of genetic variability available to the breeder.

It sounds almost too easy, but there are a great number of problems. Though botanists can get a whole plant from a single tobacco or carrot cell, they cannot readily do this with soya beans or wheat. In addition, they can fuse a soya bean cell with a wheat cell and get the hybrid cell to proliferate and divide very well in culture, but they still cannot get a whole plant. The techniques of tobacco regeneration depend entirely upon the manipulation of hormone levels in the culture medium, but the same manipulations have not been so successful for other plants.

The situation is very tantalizing and irritating. Scientists can make lovely cultures with millions of cells, for these other plants, and from these cells they can select mutants which have most fascinating qualities, but no matter how they adjust the hormone levels they cannot get the whole plant back. For instance, there is a bean mutant which overproduces one nutritionally important amino acid. Could they increase the nutritional quality of the bean crop by getting whole plants from this variety? No one knows. If scientists don't solve that problem they will still be doing mere intellectual exercises. But Peter told me recently that there are first indications that this problem will be shortly overcome in crop cereals.

The determining factor is not the particular tissue from which the plant comes. For tobacco plants it makes no difference if a cell comes from the leaf, the stem, the flower or anywhere else, though some tissues seem to work better than others. In other plants—corn, wheat, barley and potato —whole plants can be regenerated if the tissue cells have been in culture for only short periods of time. After all, anyone who takes cuttings of plants for the garden does *that* kind of manipulation all the time. But the fused or genetically manipulated cells that have been in culture for longer periods, whether diploid or haploid, seem to lose their ability to regenerate.

Part of the problem, Peter Carlson thinks, is that scientists have been deceived by the simplicity of regeneration in tobacco and carrot plants. These species worked so easily that scientists have been deluded into thinking that there is no problem at all: that it is just a question of tossing in a few hormones here and there. But the problem is really one of developmental biology, of learning just how to switch genes on and off to get development started. Consequently, scientists may have to develop other techniques. Instead of trying to grow a plant which is half soya bean and half barley, perhaps they could fuse the cells, make the hybrid and learn to drop off unwanted chromosomes by biochemical techniques. In this way most of the genetic information of one of the species could be lost, except for certain small bits with important qualities. The hybrid would be allowed to rest on the genetic background of the other species, which should then be able to regenerate more easily.

Or perhaps recombinant DNA technology could provide the answer: using a bacterium to pick up the favored gene from one plant and deliver it to another—then hoping that it will remain there. Scientists have been trying to do this with a virus which replicates very easily in plant cells. It seems to have one genetic site with information that says, "Replicate me." They have split its DNA, spliced it into a plasmid and inserted the recombined DNA into *E. coli.* They are now trying to infect mustard plants with the hybrid virus-plasmid.

Since the virus already has the information to replicate itself in other cells, and all the scientists did was attach a piece of foreign DNA to it, the technique should work. But Peter says the results have been very "Uh, uh ... Neither eureka, nor totally down in the dumps." That is the usual outcome of an experiment in science. In any case his work has temporarily ceased, due to the new safety guidelines for this research. He does not yet have the appropriate facilities.

If genetic variability is one of the main agricultural problems, the high cost of fertilizers is another. Since the formation of OPEC and the oil cartel, not only the price of oil but the price of fertilizers has skyrocketed. In the last eighteen months the price of nitrogen—a key element in ammonia fertilizer—has gone up fourfold. This is hardest on the poor countries of the Third World, but it has had its impact on richer countries too, particularly affecting the price of grain and animal foodstuffs. It makes the search for a viable alternative to chemical fertilizers as economically important as alternative sources of energy, and once again it is an area in which plant geneticists are busily at work.

The majority of plants rely on waste material that has been returned to the soil in one form or another to break down into the nitrogen they need via ammonia, the crucial compound. One group of plants, however, has for a long time coped with the problem of fertilizers on its own by entering into a symbiotic relationship with certain types of bacteria and taking advantage of the bacteria's capacity to "fix"—trap —free nitrogen from the air. These are plants of the legume family—peas, beans and clovers—and this ability is why, after a period of cereal crops, a farmer will set a field to clover for a year in order to let the bacteria restore the ammonia that has been taken out of the earth.

Scientists around the world are now working to extend this natural ability to other plants besides legumes, especially to those plants which are agriculturally important. There are several ways of going about it, and at the Nitrogen Fixation Unit at the University of Sussex, England, one crucial first step has already been taken.

The bacteria which are the slaves in this process are of two kinds: those that live freely in the soil and those which associate in the root cell of plants. The associated bacteria are the most efficient machines for fixing the nitrogen, but at Sussex, scientists are studying the free-living bacteria first,

because even though they are not very important agriculturally, they are the simplest of the systems—and the primary
step is to understand how they perform this remarkable
reaction.

Besides the chemists studying the reactions of nitrogen
fixation, biochemists are also studying the structure and
function of the crucial enzyme which converts the nitrogen
into ammonia—an essential step in making fertilizer. This
enzyme is present in both the free-living and the associated
bacteria and is the keystone of the whole process. It is a very
complicated substance, made up of at least three separate
proteins which must act in concert to carry out the reaction,
and to do it at all must consume very large quantities of
energy. It is also destroyed by oxygen in a matter of seconds,
so the scientists have had to develop some very elaborate
biochemical techniques for doing their work in an oxygen-
free atmosphere. Since researchers have to breathe, the manipulations are done in sealed bottles and through black
boxes, and this is in addition to all the precautions for keeping *out* viruses as well as keeping *in* potentially dangerous
recombinant molecules.

The process represents a slight biological paradox: nitrogen fixation is a reaction which requires enormous amounts
of energy, but it will only take place in an atmosphere devoid
of oxygen—yet it is the oxygen which by respiration gives
cells the very energy required for the process. It is suspected
that the symbiotic relationship between the plants and bacteria came about at a very crucial evolutionary stage, probably at a time when the atmosphere had no oxygen. Some
scientists at Sussex, specifically Dr. Christina Kennedy, my
friend who speculated on the new forms of life she might
make, are studying the bacteria's genes in an attempt to find
out where the nitrogen-fixing ones are located with respect
to the others, how they function as a group, and what the
various conditions are which will set them going in the first
place. If they can understand all that, then they will have a
better idea of the feasibility of asking these genes to work in
new plant cell environments or of finding ways to manipulate the bacteria so that they will be able to form associations
with other plants.

A twenty-eight-year-old scientist at Sussex, Dr. Ray Dixon, is also working on the problem by taking bacteria and viruses which are not nitrogen-fixing and seeing if he can attach a nitrogen-fixation gene to them. He succinctly summed up the prospect for me with the words "In theory any old plant—and at the moment no plant whatever!" Nevertheless, on March 18, 1976, his group reported a very crucial first step in *Nature.* Their ultimate aim is to take nitrogen-fixation genes and stick them onto a natural plant virus, such as cauliflower mosaic virus. They would then infect the plant cultures and examine what had happened to the recombinant virus molecules. The useful fact about the cauliflower mosaic virus is that it doesn't completely obliterate a plant cell as do most viruses, but it acts in relationship to it like a wart virus does to an animal cell: it replicates and survives and the cell also continues to live. But here again, Ray insists, their biggest problem will be to overcome the problem of gene expression. The complicated stages by which the instructions of a gene result in the formation of a protein are so vastly different in bacteria than in the cells of higher organisms that it may be the new environment will not permit all—or any—of these stages to take place.

The crucial genes are called the Nif genes, which stands for nitrogen fixation. The group at Sussex has now extracted these genes from a free-living bacterium that fixes nitrogen naturally and has transferred them to many different bacteria. In order to do this, they needed a plasmid and used one which belongs to a group known as "the promiscuous plasmids" because of their ability to infect many different types of bacteria. The steps are, as usual, extremely complicated technically and very ingenious, the nitrogen-fixation genes being picked up out of the bacteria from one type of plasmid and then being recombined with the promiscuous ones, which then infect the new bacteria. The detailed sequences and exchanges conjure up for me a genetic square dance, with partners being interchanged, shifted sideways and along, as pieces of material are swapped here and there. And because these promiscuous plasmids have a tremendous host range, they can be transferred to a large number of one particular type of bacteria, and the expression of nitrogen-fixation genes can be studied in a variety of settings.

Yet another possible way to solve the nitrogen problem is to modify the bacteria directly and get them to do the real job needed. Dr. Ray Valentine of California has shown that by forced mutations, certain nitrogen-fixation bacteria can be forced to excrete ammonia directly. They are fed on carbon salts and a little bit of glutonate, and can eventually be turned into little ammonia-excreting factories. Now the bacteria themselves become a cheap source of fertilizer because of the ammonia they produce. Many scientists, Ray Dixon included, think that it will turn out to be more feasible to increase the range of the bacteria or mutate them than to give them new genes, since the technical and scientific difficulties are so enormous. Here is one possible solution of great simplicity: just put free-living bacteria into a large pond and use them as a source of fertilizer. Some simple plants, such as blue-green algae, fix nitrogen and also get their energy from the sun, so they would not require feeding with much of anything. They do consume a tremendous amount of carbon in order to do this fixing, but you could put them in big ponds and utilize the sun's energy directly to enable them to produce their own food by photosynthesis. The ponds would only require the minimum kind of maintenance—a few nutrient salts, for example—and the fertilizer that is produced could simply be scooped up as required. This may turn out to be more viable economically.

The application of genetic engineering to agriculture is promising, tantalizing, and both short and long term. Ray Dixon estimates ten years at least before scientists have produced a new nitrogen-fixation plant, but in the interim they might be able to do something about the bacteria which already fix nitrogen naturally, and this by itself would be useful in terms of crop production.

It is clear that the technique of developing new hybrid forms of cells in tissue culture will not be a panacea for all the problems of agriculture, but if people like Peter Carlson can supply a wide range of germ plasm, and thus genetic variability, to those who are breeding seed crops, the future outlook, both for the world's food problem and for continuing an efficient agriculture remains reasonably bright. Genetic engineering will enable us to change the architecture of plants a little at a time, introducing new, small portions of genetic material here and there. This is a much more likely outcome, Peter insists, than producing a soya bean–barley plant, or for that matter, the duck with the orange sauce. But even in order to do this, we are going to have to learn much more clearly what the physiological and bio-chemical underpinnings of those economically important properties of plants are.

The technology is with us in theory, and given a large amount of capital, it could be developed appropriately and efficiently. In America and other countries, the large agricultural chemical companies that for the past twenty years have produced insecticides and pesticides are realizing that the era of chemical solutions is coming to an end. Insects can be killed. Weeds can be killed. Plants can be encouraged to grow big or small by putting hormones in the form of growth regulators onto them. In the long run, however, the real answer is a new agriculture, biologically developed, and the one way biology can be modified is through genetics. That is why the big companies—Du Pont, Ciba, Geigy in America, I.C.I. in the United Kingdom, Monsanto, Sandoz in Switzerland—are moving into the business of cell biology and recombinant DNA. They are buying up seed companies and establishing plant genetic laboratories within them; they are utilizing the existing expertise in microbiology and in the production of drugs by fermentation—which have produced such important antibiotics as penicillin—and are whole-heartedly transferring that technology to plant cells. It may take a decade or more, but the very same expertise that produced antibiotics and insulin will be very helpful in asking a plant cell to make more of a given nutrient, or in making a plant resistant to stress conditions, such as extremely high or low temperatures or drought.

Industry is already sniffing out the commercial possibilities of this new technology in animal products, too, as the presence of research representatives from Roche, Merck and other big drug companies at Asilomar testified. I.C.I. in England is underwriting some of Dr. Kenneth Murray's research at the University of Edinburgh, with an eye to a new commercial production of steroids and growth hormones. Du Pont and General Electric are very active in exploring the tantalizing prospect of using the bugs to eat up oil spills. The

headline in the *London Times* reporting I.C.I.'s new investment in recombinant DNA research caught the industrial
mood well, even if the prospects were a little exaggerated:
"Microbes that eat dirt and refine gold."

But all this does have a negative side, and ethical and
moral issues emerge. Once the companies have these techniques, they can patent them and their products. That by
itself does not matter, but when they have developed seeds
to respond to their drugs and substances, they could once
again form an agricultural monopoly. The kinds of seeds
they will make will be those that need their pesticides and
their growth regulators, and if such a new monopoly in this
area is developed, it will affect farmers not only in America
and Europe, but in the Third World countries, too. Since
most farmers buy seeds from one seed company, they use
their hybrid seeds, *their* hybrid wheat, barley, soya beans
and corn, and so farmers become more and more dependent
on a few companies for the source of their superior seeds. If,
then, you have an agriculture in which your seed merchant
is also your pesticide producer and your fertilizer producer,
not only will many small people who are presently developing interesting techniques and important products be forced
out, but there again may be the kind of intense concentration
and preoccupation with one method which would force agriculture down a rigid road. This has happened with artificial
fertilizers. Artificial fertilizers seemed to be a godsend, to
the Third World especially, and an enormous amount of
money and commitment went into them. It is quite clear
now, however, that not only are artificial fertilizers very
expensive, but that many of them actually repress the biological processes of nitrogen fixation, and in the end many do
more harm than good. This is especially so when farmers put
too much on the land, as they are often tempted to do. In
addition, they cause salts to leak out of the soil and make it
more liable to erosion. Thus, in the Third World countries,
where for ages the "honorable night-soil" was a standard
fertilizer, the value of indiscriminate application of chemical fertilizers remains questionable.

Genetic engineering in plant biology, then, is not going to
do everything and is certainly not going to do it in a hurry.
But one should not be tempted to deduce that because so

little is known and the problems are so complicated and the work so long term, nothing is going to happen. The history of science gives us some awful warnings: we have sometimes taken refuge in a pessimistic "death wish" in the complexities of nature, hoping, perhaps, that difficulties will save us from the consequences of unwished-for technologies. But as I talked to people working in recombinant DNA I underwent a conversion and came to believe that in the end a scientist who has said he is going to be able to do something almost certainly will.

4
Creating New Men?

"I won't eat people. I won't eat people. Eating people
is wrong."

—Flanders and Swann
"The Cannibal's Song,"
from *At the Drop of a Hat*

The island of São Miguel is the largest in the Azores, that volcanic outcrop of rocky islands lying on the Mid-Atlantic fault stretching southwards from Iceland. A small village, Bretanha, rests on the tip of the northwest corner, and from this village, as from so many others in the Azores and on the Portuguese mainland, a steady stream of emigrants has gone westwards to the United States. Most of the sailors have ended up the farthest away, in California, but the whaling fishermen of the Azores settled in places which evoked familiar echoes, such as the whaling towns of Massachusetts, like New Bedford. In the late nineteenth and early twentieth centuries, a number of peasants left Bretanha, all the children of one William Machado. Now, in 1976, many of his descendants live in Fall River near Boston. The family carries an appalling hereditary disability, however, a disease which does not begin to show until early middle age, by which time the decision to have children has already been taken. Though from time to time it may skip a generation, it is on the average a dominant disorder: that is, only one parent need carry the gene for the disease to be expressed in the children. Each child stands a fifty-fifty chance of developing it.

The disease progresses slowly and shows itself as a complete loss of motor control. It affects the deep structure of the nerves, and the speed of nerve conduction is sharply reduced. Movements cannot be coordinated. First, the victims begin to stumble and fall about; then they need to use walking sticks for support, and finally they end up in a wheelchair. Speech becomes slurred and indistinct, taking on the thickened tones characteristic of patients after strokes. There may be problems in swallowing, so the secretions of the mouth eventually accumulate and then the secondary complications of pneumonia begin. Since there is no way of diagnosing the disease in its early stages, nor any way to identify the disease by examination of the cells in the fetus, having children at all turns out to be a kind of genetic Russian roulette.

In 1845, a Portuguese whaling ship called into San Francisco and Antone Joseph, a sailor from another island, Flores, jumped ship. Like many people before and since, he was enchanted by San Francisco and California, returned to the Azores to fetch his family, and then settled in the northern part of the state, joining the miners of the gold rush. Antone Joseph also carried a similarly dominant gene inherited from his father. It produces the same symptoms as above, and was labeled Joseph's disease by Dr. William Nyham of the Oakland Children's Hospital in San Francisco. It is undoubtedly the same disease, but according to Dr. Bernard Davis of Harvard Medical School, until I came to research this book, the two groups of doctors in California and Massachusetts had been unaware each other.

Antone Joseph died of the disease. His first daughter, Mary, had fourteen children, twelve of whom lived to adult life; eight of them developed the disease and some of them passed it on to their children. In a remarkable gathering on September 29, 1975, a hundred descendants of Antone Joseph held a somewhat macabre family reunion in the auditorium of the Oakland Children's Hospital to receive counseling on their heredity problem.

The impetus for the assembly came because one of Joseph's descendants, Mrs. Rosemary Silva of Livermore, California, read an article in the *Ladies' Home Journal* about a family who had been consulting with Drs. Nyham and Rosenberg. The description of their disease sounded so like the one that haunted her family that she wrote to the National Genetics Foundation, asking them to arrange a counseling session, and with their help, traced a hundred family members over a six-month period. The meeting revealed a common substructure to their lives; at last they now know the real nature of their problem. One family group, Mrs. Silva's, had been told that their father had syphilis. The relatives never wanted to discuss the disease, and the children were soothingly told that they could never get it. But as Mrs. Silva said, "It is hard to hide when you have relatives dying everywhere before their fortieth birthdays." Another woman, Mrs. Doris Newbaurm, said she was tired of having the cops pick her up, thinking she was a drunk or a drug addict, and spoke poignantly of the time she visited her mother, who was sick with the disease. She stumbled as she approached her mother's bedside; then she knew that she, too, had it, and her mother also knew. "I saw the look on her face," she said, "but we made a joke of it. I told her I planned to live one year longer than she did. She said, 'More power to you.' "

These people know that they do not have too much time for life, and they also know that if they get married they may pass on the family curse, but with no way of knowing for certain either way. The extent of intermarriage amongst these families remains an unknown question. It has been denied strongly by the descendants in Fall River, but the doctors and geneticists believe that there must have been a tremendous amount in order to account for the frequency of afflicted children in subsequent generations.

Exactly what the nature of the genetic defect is remains unknown. Certainly one important gene is defective, which leads to the death or deterioration of certain brain and nerve cells, but no one really knows if it involves simply the absence of a particular enzyme or something more complicated. If it is the absence of one identifiable enzyme, then ultimately there may be a hope for remedy in one individual's life at least, though it will not of course eliminate the gene from the family.

Joseph's disease is just one among many genetic diseases, the most well-known being sickle-cell anemia and hemophilia; in both cases there is a single gene defect. Another single gene disease is Lesch-Nyhan syndrome. Male children who have it start biting their fingernails, but instead of just biting their nails like ordinary children, they start to chew off their fingers. Then they attack their lips and tongues. They become psychotic and usually die before the age of fourteen. It has been tracked down to the lack of one single enzyme, called hypoxanthine-guanine phosphoribosyl transferase. Though there is no treatment for the disease at present, at least it can now be diagnosed *in utero* and an abortion offered. Other diseases result from whole chromosome abnormalities. Down's syndrome, or mongolism, results from the presence of such an extra chromosome. But of all these genetic diseases, Joseph's disease and a similar one, Huntington's chorea, probably represent the ultimate in tragedy, both because of our present incapacity to diagnose them early, and because of the agony of the parents who must play genetic dice when having children. Here at this point, if any, one would hope that science might help. But how?

Some beginnings can be made: indeed, they have been made already in the case of single enzyme defects—insulin for hereditary diabetes, for instance. But in certain other defective-gene/missing-enzyme diseases, we still are not able to supply the required substance.

It is situations and diseases like these that present geneticists and doctors with perhaps their greatest medical challenge. The notion of being able to replace defective genes by correspondingly normal genes is a prospect both dazzling and daunting—dazzling because of its scientific and medical promise, daunting because it is not simply a question of inserting the correct gene just somewhere in the body. It means getting the gene in, in the right amount, at the right stage of development, in the right cells, where it will be subject to the right kind of regulation; and that possibility is

years off. Superimposed upon the dual problem of identifying the gene in molecular terms and designing and making
a vector to carry it into the cells where it is required, is the
whole problem of the regulation of development. If this is
difficult in plants, as Peter Carlson emphasized, it is infinitely more difficult in animals.

Nevertheless, the first scientific steps to replace missing or
defective genes have already been taken, and the experiments, representing as they did one of the early dramatic
attempts at genetic engineering, made the headlines. *News-
week* (October 25, 1971) described them under the banner of
"Genetics: A Friendly Virus." *Time,* on the same day, spoke
of "Transplanting a Gene." The English *Sunday Times* ran a
front-page headline, "Repairing the Human Cell." Dr. Max
Perutz, Nobel Laureate and chairman of Britain's Medical
Research Council's Molecular Biology Laboratory in Cambridge, described it as "the first step toward therapy for con-
genital diseases." The article in *Nature,* under "News and
Views" that accompanied the first reports, was written by Dr.
John Tooze, now head of the European Molecular Biology
Organisation. He spoke of ". . . a claim little short of revolutionary." John Tooze could find nothing seriously wrong
with the design or the execution of these experiments,
though he warned that people's "minds would flood with *a
priori* skepticism and prejudice." Five years later, however,
the experiments have only recently been confirmed and are
still a matter for controversy; for a while one of the leading
investigators, Dr. Carl Merril of the National Institutes of
Health, who has an excellent scientific reputation, was in a
scientific "doghouse."

I went down to see Carl Merril in December 1975. The
grounds of NIH are delightful, especially on a warm December afternoon, when the weather gives a last spurt, making
the phrase "a dying fall" a literal rather than a metaphorical
. expression. But I find the atmosphere of NIH somewhat op-
pressive, not because of the people there, for those I have met
have been delightful, relaxed and friendly, but because in no
other place does one get such a feeling of the scientific factory: an endless range of corridors laid out in squares or
rectangles with rooms and laboratories repeated almost exactly on six floors or so, all alike and only the labels on the

doors to provide obvious variety. Designed, no doubt, for the greatest possible economy and efficiency, it is a die-stamped atmosphere, as if discoveries, too, could be banged out with a thump of a bureaucratic hand. But the scientists have learned to live with it. It is not too difficult, my friends informed me, to outwit, outmaneuver or simply ignore the bureaucracy. Such is its weight, however, that if the bureaucracy chooses, it can smother you, especially after a brilliant success that perhaps turns out not to be quite such.

Scientists don't like to be associated with suspected failure, and there is a kind of primitive fear amongst them, as if a failure might be contagious. At thirty-nine, Merril now shrugs this off. He is a man with a confirmed reputation, an honest confidence in his work, and an extremely far-ranging imagination. Of all the scientists I talked to while working on this book, only two were capable of leading me in almost any direction, into a range of infinite and fascinating possibilities. Peter Carlson was one. This somewhat intent young man with a mop of frizzy auburn hair was another.

If for the past five years he has been working in a situation with a degree of isolation, this is nothing new in his life. His father, a doctor of oral surgery at the City Hospital in Washington, D.C., acted on an unusual assumption: that it was unreasonable to charge patients when they couldn't afford to pay. Consequently the family was always very poor, and Merril grew up in a slum, a brilliant, intelligent and aggressive kid. He could well have been lost to science because, to put it mildly, he was very counter-suggestible at school—a real rebel. He and two other kids were saved, he told me, by the sympathetic imagination of a woman teacher who took the three little eight-year-olds in hand and brought them to her house, fed them and set them to read and learn. To this day Merril can't write in script; he can only print. Finally he went to Georgetown Medical School, where he qualified as a doctor before going into basic research.

What Merril and his co-workers Mark Geier and John Petricciani did was to infect defective human cells with a virus that had picked up the missing gene from bacteria. They began with a human cell line taken from a patient suffering from an hereditary disease, galactosemia. As a result of this defect, the patient is unable to utilise galactose, a simple sugar found in dairy products. Unless a baby with this disease is put on a diet free of milk and milk products, it will begin to show symptoms of mental retardation and malnutrition, and ultimately will die. The first step was to infect the bacterial cells with the virus, which then replicated itself, and new viral particles were released. These particles were then used to infect defective human cells, in culture, and the cells were then tested to detect the presence of the gene activity.

In their experiments, the scientists used cells taken from the skin of one of these patients. They also took the bacterial virus *Lambda,* which had picked up from *E. coli* the genes necessary for the utilization of this sugar. Having got the required bacterial gene incorporated into the virus, they infected cultures of the human skin cells.

Now this experiment did *not* supply a normal *human* gene to an abnormal human cell. It supplied a normal *bacterial* gene that could perform the missing function to an abnormal human cell. The genetic instructions given to the human cell were alien, but the evidence seemed to show that they were being translated, and that the key enzyme necessary for the correct utilization for galactose was increasing in the culture. If this was so, and they could indeed replicate the enzyme, then science had indeed taken a first step towards "repairing the human cell."

The criticisms started at once, and all were predictable. First, as John Tooze's article in *Nature* had indicated, the work really was revolutionary and as such could be expected to attract the full weight of critical assessment. "Everybody," *Nature* said, ". . . will be out to find flaws in their work." This was especially so because of the world-wide publicity given to the experiment, for scientists have very ambivalent feelings about press coverage of other people's work, and sometimes of their own, and generally distrust the way the media handle science. They are especially uncharitable about those experiments which, by design or default, attract the kind of highly favorable publicity that Merril's work did. The bureaucracy too is nervous under those conditions, for if the results don't hold up there is always a possibility that the publicity will backfire. In such cases the burden of proof that is already on a scientist is greatly intensified. Indeed, the work did run into difficulties: some people jumped in and couldn't reproduce the effect, and some of them were very vocal about their failure.*

But in addition, Merril and his colleagues immediately ran into a problem that continues to afflict those who work in the area of genetic engineering, and that will probably always do so. There were some people at NIH who felt, and still do, that scientific work should not go in that particular direction. Though Merril considers he was just doing basic research, trying to find out more about viral activity, others saw his experiments as the first steps towards a morally reprehensible Faustian attitude towards human life. In the case of one or two people in the animal house, the disapproval became active: they refused to help with the experiments at all.

So attacks came from two sides: expected scientific ones from those who felt that the results of the experiments were either ambiguous or unrepeatable and unexpected ethical ones, from those who felt the moral implications were just as grave. As several of them told Merril, "Man should not play God."

In 1976, five years after this experiment was first done, it doesn't really matter whether the results can be repeated in all their fine detail. Claims and counterclaims over this have recently surfaced once again. What is more to the point is the fact that, as David Jackson insisted, workers in Australia have both claimed to do that and have shown that other bacteria, and other bacterial virus genes, can be expressed in plant cells, so that infection of mammalian cells in this way is a very likely possibility. Therefore, the prospects for gene therapy of some sort are reasonable.

Instead of daily injections of insulin, for instance, the diabetic might be injected with a virus carrying the gene for the appropriate enzyme and let the virus do the work. The success of such a technique would ultimately depend on the ability of the virus both to infect sufficient cells of the body

---

*M. Rosenberg of the National Cancer Institute has just demonstrated that the messenger RNA for galactose of *E. coli can* be expressed in a human blood cell.

for the enzyme to be produced in quantity and to remain in an infective state for the duration of the patient's life—both of which are totally impossible at this point. Multiple injections of the virus may be necessary. Moreover, it may well be that, in the total human organism as opposed to a human cell culture, the immune responses will react to this virus just as they would to any other and suppress its activity.

Ideally, of course, one would like a technology that would deliver not a bacterial gene but the correct *human* gene, and allow it to substitute for, or suppress, the defective human gene in the cells. However, such a technology would do nothing at all for the problem of inheritance of the defect: it would only improve matters for one individual in his lifetime. The real question is, could one change the germ line so that, for example, the Portuguese descendants of Antone Joseph and William Machado would never carry the disability? One way, of course, is very simple: just cut it off, don't allow these people to reproduce. Elimination by chastity will always remain one answer. But could scientists develop a technology which would both detect the defect in the early stages of the fetus—when the embryo consists, for instance, of only sixteen cells, the stage before the germ line is affected —and then correct it? Such a technology would involve fertilizing the embryo in the test tube, then laboratory manipulation and then reimplantation of the embryo in the mother. Only then might one ensure a viable embryo without a defect that would be carried to future generations. To do any of this would be a formidable technical problem, and at present it is quite impossible.

Scientists have, however, in certain experiments, managed to fertilize eggs artificially in test tubes, and then implant them in the uterus of the mother. The therapeutic rationale for even starting such experiments, which brings us very close to certain basic ethical questions, has been to develop a safe and successful technique for overcoming some forms of human infertility, where, for example, because of blocked Fallopian tubes, an egg cannot enter into the uterus for fertilization. There are many such uses for such techniques. They could be a successful tool to help those women who presently cannot conceive their own children. Similarly, they could be used in surrogate motherhood: if a woman could not bear children because of a defective uterus, another woman could be asked to receive a fertilized egg and bring it to full term for her. Finally, in a situation where there are known genetic defects, the errors could perhaps be manipulated out of the chromosome, and a genetically reconstituted and healthy embryo reimplanted back into its natural mother.

Artificial fertilization of mammalian eggs by sperms in the test tube and their subsequent reimplantation in the uterus of the mother have been tried in a number of animals and successfully accomplished in cattle, rabbits and baboons. Test-tube fertilization of human eggs and sperm has also been successful, though no genetic manipulation has yet been undertaken. Successful reimplantation of such artificially fertilized eggs in the uterus of the mother and surrogate mothers has been claimed for humans, but has not been generally accepted.

In the baboon experiment, the animal mated normally and her fertilized egg was removed five days later, before it became attached to the uterine wall. It was then implanted into the womb of the surrogate baboon mother, chosen for this privilege because she had ovulated on the same day as the genetic mother. This ensured that the surrogate uterus was in the correct cyclical period for supporting an embryo.

A similar attempt in humans was first reported on April 24, 1975, in an article in *The Lancet,* by Dr. P. T. Steptoe, a surgeon, and Dr. R. A. Edwards, a physiologist. This reveals amongst other things, the extremely complicated—Peter Carlson calls them tortuous—procedures which have to be gone through before embryo reimplantation, let alone genetic engineering on humans, is possible. The authors claimed the first successful transfer of a test-tube fetus into a mother, from which a pregnancy that lasted three months was initiated. The patient in question had been successfully pregnant once, but had failed to conceive again. For six years she had tried again, partly by natural methods, partly using reimplantation techniques developed by Drs. Steptoe and Edwards. After two attempts they managed to achieve an artificial fertilization of her egg with her husband's sperm. This egg was reimplanted through the cervix but was finally expelled. After a further eighteen months of hormone treat-

ment, eggs were retrieved once again, fertilized artificially
and reimplanted, and other hormones were given to assist
the pregnancy. This treatment continued for two days before
implantation and up to seven weeks afterwards, in the pregnancy. Some ten weeks later, however, the patient complained of severe pain and it became clear that the embryo
had implanted in the Fallopian tube rather than the uterus,
and the fetus died. Finally, the right Fallopian tube was
excised altogether, and though attempts were made to culture the embryonic cells, this was not successful either.

The scientists concluded that the fetus died because of its
location in the oviduct, and suggested that the low success
rate of implantation in earlier work may well have been due
to adverse conditions in the uterus, or to the techniques
employed. They believe that treatment with hormones may
always be needed to induce implantation especially if, as in
the case of this patient, hormones were necessary in the first
place to get the eggs.

Clearly these procedures are not without their attendant
difficulties to put it mildly! Even the word "tortuous" may
be an understatement when one considers the battery of
hormonal treatment that was necessary for the human patient. After the recent cautionary warnings about the long-
term effects on the child of hormonal treatments during the
pregnancy of the mother, we may question how advisable
any of these procedures may be, given our present state of
knowledge. Besides, other problems are likely to arise. Two
recent reports (*Nature,* March 1976, and *The Lancet,* April
24, 1976) from the Clinical Research Center and University
College, London, respectively, have shown that mouse eggs
fertilized outside the body are five times as likely to have
chromosome abnormalities as those fertilized inside the
body. The defects generally took the form of tripling the sets
of chromosomes and, in nature, a fetus with three sets of
chromosomes is always aborted. Though there can be—as
yet—no evidence that these kinds of abnormalities *will*
show up in similarly treated human cells, it does mean that
the procedures will have to be studied very carefully. Manipulation of rabbit eggs, by Dr. Bromhall of Oxford, followed by reimplantation has already been shown to result in

abnormal development. Because human eggs are so very small, it may not be possible to examine each one minutely for signs of potential trouble before replacing it in the mother. It would be ironic if the procedures designed to repair a defect resulted in an equally serious but different defect. This may itself be a barrier to genetic manipulations on human embryos.

Another possibility is to bring the whole pregnancy through in a test tube and avoid damage to the fetus due to much handling. Is this a serious possibility? Some people clearly think so. Certainly the scientific challenge is dazzling, and provided the technical obstacles can be overcome, scientists will do it. Indeed, they see no reason why they should not. In a notable comment reported by Albert Rosenfeld in his book, *The Second Genesis,* one scientist said, "If I can carry a baby all the way through to birth *in vitro,* I certainly plan to do it, though obviously I am not going to succeed on the first attempt or even on the twentieth."

The scientist in question refused to be identified, and one must ask "Why?" People who have enough self-confidence— one might call it hubris—even to contemplate this kind of experiment do not tend to be shrinking violets or to shun the limelight. Such modesty must have another cause, and I suspect it is related to the furor that such an experiment might raise. With every day that passes, science is edging ever nearer to the modification of all life, including human life. This raises certain profoundly disturbing implications in the minds of those in society. There is, on the one hand, a deep distrust on the part of those who believe that too many scientists hold superconfident, omniscient, "we know what is good for the race" attitudes. There are others who are concerned about man's ability to handle the ethical and moral dilemmas that will result. Some have strong aesthetic and spiritual feelings of revulsion, sweetly summed up for me by Francis Bennett, who said simply, "Making babies in test tubes just isn't nice." Like the first cannibal who must have said, "Eating people is wrong," he may well be right. It all adds up to one thing: society is beginning to insist that concerns other than narrow scientific ones be considered—

perhaps even before such work is done. Doing an experiment
merely because you find it scientifically intriguing may be
just not good enough. It is certainly no longer self-justifying.

So now let us examine the techniques of genetic engineering that are at hand and extrapolate from them to the future,
at least so far as we are able. Then we can try to face something of the extrascientific issues—Alvin Weinberg calls
them "trans-scientific"—that will inevitably arise.

It is sometimes very difficult to get a feel for what scientists themselves really believe about prospects for the future
in genetic therapy and various kinds of genetic engineering,
over and above the simplest kinds that nature and man, with
his selective breeding techniques, have been engaged in for
millions of years.

Whenever I have talked to scientists about this, I have
found that their attitudes seem to be governed partly by
whether they think their interlocuters are for or against
science. If you seem to be a person who, on the whole, is
prepared to give the scientific enterprise encouragement,
blessing and tacit approval, then you get an optimistic reaction for genetic engineering, though always with the riders
of: "difficult technological problems to be overcome . . . many
years away . . ." and so forth. If, however, you seem to be
against science, then the scientists will retreat a great dis-
tance and emphasize that these kinds of applications are so
far in the future, some thirty or fifty years at least, and so
highly complex, that they are really irrelevant to the issues
which are presently important, such as the safety of recombinant DNA. Thus, they seem to say, these ethical difficulties
should neither enter into our considerations at this stage, nor
be used as an argument to slow down the pace of scientific
research.

But their public utterances speak for themselves. Scientists are divided. For example, even in 1972, before recombinant techniques got off the ground, a distinguished cell
biologist, Dr. James Danielli, director of the Center for Theoretical Biology at the State University of New York at
Buffalo, emphasized that "soon, twenty to thirty years, but
we may well be there in ten years," scientists will be able
"to create new species" and "carry out the equivalent of

10,000,000,000 years of evolution in one year." He noted that it would be necessary to develop certain biomolecular techniques which would have to include not only bringing new genes into a cell, but bringing in the control mechanisms of the cell, too. But, he blithely continued, "none of these problems appear to be of exceptional difficulty"! Dr. James Watson as well has expressed public beliefs on this matter. He wrote a famous article on cloning man, in the *Atlantic Monthly,* in which he urged that society face that possibility. Dr. Joshua Ledeberg has also written extensively about the promise of biochemical and genetic engineering. It is, as Dr. Norton Zinder of the Rockefeller University has said, just on the horizon.

The New York Academy of Science recently (1975) held a conference on the whole problem of recombinant DNA and genetic engineering. In the course of a wide-ranging paper that covered the current therapies for genetic diseases, Dr. Theodore Friedmann of the School of Medicine at the University of California, San Diego, said:

> Until two or three years ago, examinations of the prospect of gene therapy for human genetic disease took a certain comfort from the degree of remoteness and unreality that characterized the subject. As the organizers of this conference, as well as many of the participants, have come to realize, whatever complacency we all felt about the likelihood for the development for the techniques of gene therapy has now been, or is now being, shattered.

In two important addresses given in 1968 and 1970, Dr. Robert L. Sinsheimer spoke glowingly of man extending the limits of his own brain by understanding and intervention:

> We can begin to confront chance and choice: soon we shall have the power consciously to alter our inheritance, our very nature.... The great discoveries in genetics and the great discoveries yet to come open a new dimension of human potential, a new route for the improvement of man.... We, mankind, are to have the opportunity to design the future of life, to apply intelligence to evolution.... What an outstanding chance and infinite challenge.

This was heady wine indeed, but it is interesting to note that since then, Dr. Sinsheimer has reversed his position: not about what is possible but whether we should do it. And it is the implications of recombinant DNA that have made him change sides.

The engineering of intelligence or beauty is, according to Dr. John O'Brien, also of the University of California, San Diego, "like going to the star Alpha Centauri. It is technically feasible: we know we can go there if we want to spend the time or money." The time factor, though, will be on the order of generations.

To close this circle a little, and to bring us back to earth with a bump, we should notice, however, that Sir MacFarlane Burnet, Nobel Prize winner, commenting on the possibility that Carl Merril felt he had achieved, a virus that would deliver a replacement gene that would then function in a foreign cell, said: "I am willing to state that the chance of doing this will remain infinitely small to the last syllable of recorded time."

When scientists differ amongst themselves, as they do over the issues raised both by recombinant DNA and genetic engineering, what, I ask plaintively, is the poor old amateur supposed to believe? I remember two things. When I first interviewed Peter Carlson about this book, he was quite pessimistic, feeling that plant biologists were blocked because of their inability to grow a whole plant back from a fused hybrid cell. Only nine months later he was thoroughly optimistic: the problem would eventually be past history, such was the promising pace of the work. Second, Mac Burnet's words evoke in me, an historian, strong echoes of *déja vu.* There are awesome but piquant precedents for his words in the history of science. Was it not Rutherford who said that "No practical result will ever come from the splitting of the atom"? Was it not Sir Richard Woolley, British Astronomer Royal, who said—and even said it to the Russians—"The prospect of travel to outer space is utter bilge," thus evoking a memorable cartoon in *Punch* two years later, portraying the Russians showing the Astronomer Royal their space station? The caption read: "And now, Sir Richard, we invite you to take a look at our prospects for utter bilge."

I think we would be wise to hedge our bets and assume that eventually scientists will do what they say they think they can do, in ten years from now (pace Dr. Danielli), or

thirty years from now (pace Dr. David Jackson), or generations from now (pace Dr. John O'Brien).

Thus there is a great spectrum of possibilities for the future, ranging from a gene that can make its protein in an
incompatible and totally foreign environment, to the engineering of desirable traits in organisms, including humans;
from a small experiment in a minor laboratory of NIH to the
equivalent of a biological space trip to Alpha Centauri. In
fairness to the scientists, however, before we extrapolate
well into the future and examine some of the problems, we
must once again distinguish between what is and what is
not, between established fact and fantasy—imaginative, possible fantasy, but fantasy nevertheless. As Dr. Charles Weissmann told me: "What we would like to correct, and *could*
correct, are two quite different things. Given a disease that
comes about because one gene is missing, well, we *may* be
able to introduce that gene. But even that is already asking
a lot. But where multiple chromosomes are concerned,
chromosomes that have to be dropped off before the cell will
function normally, or where the control mechanisms are
messed up, it is all far more abstract. Gene expression and
gene control is still a big mystery. So to do genetic engineering with any subtle and sophisticated biological property,
like intelligence, or with one which depends on the cooper-
ation of many genes—well, we just can't. For we don't really
understand *one single gene*, yet." He smiled deprecatingly
and went on: "But for that, however, *all* the techniques are
available. In three or four—at the most five—years, we *will*
get the genes of higher organisms expressed in bacteria. We
may even synthesize these genes and get them expressed."
It took much less than four years. We spoke in Zurich in
July 1976, and within one month (August 29, 1976) it was
announced that a team led by Dr. Khorana of MIT had done
just that—a culmination of nine years' effort. Using chemicals pulled off the shelf, his team had already, years before,
synthesized a gene artificially. Their present one, however,
when placed in a bacterial cell, actually functioned, and the
method they have developed makes it possible to alter their
gene at any one point in the sequence of 199 components. So,
step by step, they will be able to analyze its functioning and

find out just how the signals for stopping and starting gene activity work, how the gene is controlled and regulated. It is the beginning of the end of the "big mystery"; it is the first small steps toward genetic manipulation. If understanding is the essential prerequisite for genetic control, then the floodgates of knowledge have already opened.

As I spoke with the scientists, it seemed to me that if we wanted to look at the ethical consequences of future actions in this area, we actually didn't have to extrapolate too much. So much technical mastery is both already to hand and being applied to animals as well as plants. The total spectrum of human genetic engineering may not yet be with us, but the specter most certainly is.

Would you like a child with four parents? A living mouse created by a woman scientist, Dr. Beatrice Mintz, fits that specification. Dr. Mintz took two mouse embryos when the fetuses were just a few cell divisions old. She placed them both in a dish, dissolved the protective membranes around the embryos so that the cells could amalgamate, and inserted the composite embryo into a surrogate mother, who finally gave birth. Actually the mouse had five parents, depending on one's definition of motherhood: four genetic and one for pregnancy. Why did Dr. Mintz do this experiment? Because she is fascinated by the most challenging problem of differentiation: how does a complex organism develop, with cells differentiated for separate functions, from a single cell? Since the parents had different pigments in their cells, Dr. Mintz's technique will allow her to trace the fate of individual cells through the myriad divisions the fetus undergoes. Fascinating technical applications can also be made to follow from this strictly intellectual interest. Would the members of a commune perhaps wish to extend the notion of common parenthood and responsibility to this extent?

Or would two parents like two identical children, or even two hundred? The same techniques of cloning described by Peter Carlson for plants are being applied to animal cells. What amateurs have been doing for years, when they take advantage of the capacity of plants to reproduce from root cuttings, is now being done by scientists. It was in the fifties that scientists first sliced up carrots and suspended them in slowly revolving flasks so that free cells could be released.

From these, new adult carrots, carbon copies of the first, were grown.

Then Drs. King and Gurdon of Oxford made carbon copies of frogs. The process was a little more "tortuous," but it finally worked! They took the nucleus from an unfertilized frog's egg and replaced it with another from a culture of cells from an adult frog. Then the egg was stimulated artificially —a needle prick is all it takes—and it began to develop. An adult did not grow immediately. The nuclear transplantation had to be repeated through a whole series of unfertilized eggs, but eventually adult frogs were produced—from one single cell in culture.

Then Dr. Bromhall, also of Oxford, tried with rabbits and had such success that the *London Times,* a paper not usually associated with sensational headlines, carried one (February 2, 1976) that read: "Production of Identical Humans Is Nearer." If frogs were more difficult than carrots, rabbits were much, much worse than frogs. A frog's egg can be seen with the naked eye, as any small child who wants to grow tadpoles in the bathtub will tell you. But the eggs that Dr. Bromhall used were only a few thousandths of an inch across. He took unfertilized eggs as before, but made no attempt to extract the nucleus from that minute speck; he merely introduced additional ones. Though he did not manage to do this with 100 percent success, he was successful several times, and some of the eggs even went on to divide as though they had been fertilized normally. The usually staid newspaper was perhaps a little ahead of the circumstances, however. Admittedly, the *Times* said only that the day for producing identical humans was "nearer," not that it was nigh, but though Dr. Bromhall expects that his transplanted eggs *did* contain the correct genetic instructions needed to clone another adult rabbit, he will not know this until he can introduce it into a surrogate mother and bring the embryo to full term. This he has not yet done and may never do, for without exception, even when he was successful in introducing an additional nucleus into the egg, which then went on to divide, the embryo showed gross abnormalities the moment differentiation began. Once again, however, it may be only a matter of time and technique before this problem is also overcome.

So perhaps Dr. George Wald's nightmare can come true. He is at Grand Central Station and sees eight Albert Einsteins buying eight identical copies of the *New York Times*. The *London Times* might be more appropriate! My nightmare is slightly different. I see the eight *unidentical* surrogate mothers of these Albert Einsteins, all saying, "My son, the scientist!" Surrogate mothers will be necessary for the successful cloning of human beings, indeed, of any mammal, whether it is for parents who want two hundred similar children, or for a dictator who wants two hundred thousand similar soldiers. Even if that number of surrogate mothers could be found willing to act as milch cows, we may be thankful that it will take just as long for those two hundred thousand soldiers to grow up as it will take for one soldier.

Dr. Bromhall's interest that led to this extraordinary experiment is the same that led to Dr. Mintz's experiment: the desire to find out more about development. Similarly, the embryo transplantation experiments—and this is what they are, even if conducted in humans—of Drs. Steptoe and Edwards were prompted by a desire to help those women who were failing to conceive. It was these women who sought the help from the surgeon-physiologist team. Yet once again, basic understanding lays open the possibility of frightening full-scale exploitation, and sometimes the two seem much too close for comfort.

Consider, for example, those weird hybrid organisms I met earlier on, those that could result from the fusion of cells from different species. When Dr. Hayden Coon of NIH fuses a hamster cell with that of a human, he does so because he is genuinely interested in finding out the details of the genetic control mechanisms in cells. Nor is he worried about a macabre applied technology resulting from this, since at the moment it is unlikely that his hybrids will grow. They generally proceed to drop off chromosomes, one by one, leaving the hybrid cell so functionally defective that it could not develop. This way, piece by piece, by a process of functional elimination, Dr. Coon can map the entire spectrum of mammalian chromosomes. But it is only a matter of time before our knowledge of these control mechanisms is so detailed that we shall be able to alter and modify them at will and grow these "things." We intend to do this with plants, where fusion of hybrid cells, such as those of tomatoes and potatoes, is being done for good agronomic reasons. It has been suggested that a cow-antelope cellular fusion, resulting in a "cantelope"—not a cantaloupe—might provide an ideal grazing animal for Central Africa. But what of humans and gorillas, or the Japanese woman that intends to mate with a gorilla: in all these instances I would never get any answer to my questions "Why do you want to do it?" "What do you hope to learn?" other than it would be "scientifically intriguing." The latter case was a rank commercial enterprise, just ripe for sensationalism.

As a civilization we have not outgrown what to me is a rather sickening interest in the macabre. People used to flock to the circus to see the fattest woman on earth or the dwarf or the monster or even the Dionne quintuplets. Such aberrancies pull in the audience and the cash. So, too, we may confidently expect that commercial interests will eagerly exploit the result of such bizarre unions. People in earlier periods of history never bothered to ask questions about the feelings of the exhibited, nor may people now. Yet as the swirl of controversy around genetic engineering intensifies, society is beginning to ask questions about the ends to which all this experimentation will lead as well as about the motives that directed the scientists in the first place. (This issue will be considered at greater length in Part II.)

What are the issues and the ethical problems raised by wholescale genetic engineering? The first very simple one was emphasized by Dr. Friedmann at the conference mentioned earlier. He drew attention to the need to reconcile "the rush toward scientific advance with the health needs of the sick patient." He said, "There is a major danger of becoming so seduced by beautiful new science that the primary obligation to the health of the sick patient is in danger of coming into conflict with, and becoming subordinate to, the alluring scientific work." The issue initially arose over cancer research, and though many basic scientists, of whom Dr. James Watson has possibly been the most vocal, have emphasized not only the need for such research, but the length of time it is going to take before it will be effective, this in no way reduces society's obligation to those who are presently suffering from cancer. So, too, with genetic diseases: we have a moral obligation to work toward more effective therapies for those who are presently suffering, as well as toward supplying the basic research for the applications that are still in the future.

That is one kind of priority—but there is another. Which of the genetic diseases demand our first attention, and on what criteria? Should we try to reduce the incidence of genetic disease in the population, or should we do something along eugenic lines to deliberately improve the human gene pool? Should we do everything possible to preserve genetic diversity in the species, even if this means permitting harmful genes like the one that produces hemophilia to remain? This diversity may be a very fundamental prerequisite for variation, and therefore genetic vigor.

We must neither make any *absolute* judgments about genetic defects, nor be in too much of a hurry to engineer them. Some characteristics may be both useful and harmful at the same time. We might well wonder why during evolution the gene for sickle-cell anemia, so prevalent amongst the black population, has been retained. It is so obviously harmful that one would expect it to have been eliminated by natural selection. There is an advantage associated with this condition, however, for those who carry this factor have been found to have a high resistance to malaria.

Genetic disease must be seen in its correct perspective, and in the total context of human disease. Medical research in general, as practiced in the developed countries, is largely directed to the problems of the developed countries: cancer, old age, overeating. Though it will seem a cruel thing to say, genetic diseases, taken on the world stage, are very rare: they affect only 12 percent of the world's population. By comparison, the figures for parasitic diseases are enormous. Schistosomiasis afflicts 250 million people. The population at risk from malaria is 1.5 billion. If one uses the world medical situation as an argument for genetic engineering, the argument dictates that scientists presently working on genes or gene therapies would be better employed developing a vaccine for malaria or one for schistosomiasis. There is none for either disease. We must also consider the problem in economic terms. It may well be that the application of technology to fertilize, implant and successfully develop an embryo in a mother by artificial means will ultimately go the same road as heart transplants. It will cost 50,000 dollars a go, and will not be very successful at that! Economically, embryo implantation may simply not be worth the time and effort. Adoption may be not only the most reasonable alternative but, given the overpopulation problem, the most socially acceptable too.

Another question: should one apply the same kind of vigorous life-saving and life-supporting technology to fetuses that in earlier centuries would have died because of chromosome aberrancies, but which because of our biomedical technology we can now keep alive—even though they may be prematurely born and suffering from gross defects? Other issues will arise as well, when things go wrong as a result of our manipulations, as they undoubtedly will from time to time. Suppose the mother of a test-tube fertilized and implanted child gives birth to a genetically defective offspring? Will the scientists concerned feel that they have had any responsibility in this matter? Will they feel more, or less, responsible than those who worked on the development of thalidomide? And if the fetus is found to be abnormal, whose decision will it be to terminate its life?

Professor Roy Schmickel of the Department of Pediatrics and Genetics at the University of Michigan, who himself is working on recombinant DNA as a tool to improve our understanding of human genetics, feels that it is not right to risk an operation on the fetus—whether to repair a genetic defect or a structural defect—until you know for certain that it will succeed, and this we cannot guarantee. It is quite unethical, in his view, to subject the child to a life of difficulty or sorrow because of the very condition for which one might have aborted it in the first place. If there is a risk of 50 or even 25 percent that there will be difficulties or abnormalities, then Professor Schmickel feels it is best to start all over again. What some people called "neonatal medicine" is not medicine as it has been classically practiced—that is, action in the best interests of the patient—so much as experimentation in keeping a viable fetus alive. As such, it is not much different from keeping people alive at the other end of a life-span. Professor Schmickel believes the phrase "quality of life" must be applied in *all* situations, and allowed to overrule the question of existence both after birth and before death. Certainly a lot of medical research *has* helped many babies who were a little abnormal into complete normality. His viewpoint is not against fetal technology as such, but against its too-vigorous practice. If, when weighing as little as 400 grams, the fetus cannot stay alive except by heroic measures, then he believes that it is wrong to impose a likely handicap on that child by doing everything we can to maintain a life. Most premature births occur because something has gone wrong, and in the past, weak or defective babies often died, for abnormalities result in natural abortions. By applying heroic measures in neonatal periods, we may certainly improve the infant mortality figures, but we may be defeating our own goal—that of having strong, healthy children.

Surrogate mothers raise other issues: are we justified—is *any* human being justified—in asking one woman to go through pregnancy, with all its physical and emotional commitment, on behalf of another woman, only to have the child taken away at birth—even at a price? It has been done. Since his wife was infertile, a man recently paid $7,000 for a surrogate mother, who was artificially impregnated by him. Are we happy with the new concept of "Wombs for Rent"? Shall we regard this as no more than a twentieth-century extension of the medieval idea of the wet nurse? Are we willing to contemplate a situation where paying one set of women to bear children on behalf of others is regarded as a reasonable way of solving poverty and unemployment? I have had this put to me as a serious proposition several times. Would this not lead to the same set of problems that already arise when blood donations are paid for, as they are in the United States? It will be the underprivileged, the poor and the deprived—those who may not be in the best physical shape—who will be obliged to take this kind of work.

Yet another scientist, Dr. Bernard Davis, put the opposite view to me in characteristically vigorous terms. "I have no interest in having our advances in medical science change our general social patterns to the degree of busy women hiring others to be their cows to produce their offspring." I applaud his views. Too busy to have the children may also mean too busy to look after them.

As for the scientist who hopes to grow the baby in the test tube: whose child would it be? Will he be willing to undertake the same kind of loving responsibility for that life with the care and attention that is not only the right of every human individual, and which is essential for normal, happy growth, or will it just be his "experiment"? If the child is defective in some way, would he be willing to undertake the responsibility for its care, or spotting this at an earlier stage, would he eradicate this life as simply as he began it?

To conclude: genetic engineering is nothing new in history. As practiced by man it has been going on ever since life appeared on this earth. Peter Carlson told me he considered that the Stone Age marked "the finest hour" of genetic engineering, and he is not being facetious. By sowing and reaping in a thoroughly selective manner, crop plants were developed that were well suited to man's new agriculture, and they would never have evolved at all but for man's intervention. The development of such an agriculture marked the beginning of civilized periods as we have come to know them. If as a result of the new techniques of recombinant DNA, along with other contemporary scientific sleight of hand, we are now able to extend to all organisms—man included—the same kind of manipulations we have done for centuries on plants, some will not only say, "So what's new?" but will also say, "Why ever not?"

Before answering that question, we must first notice three important distinctions between genetic engineering as traditionally practiced in nature, and as presently practiced by man. When widely divergent genetic materials were combined and recombined in nature, the resulting hybrid *had* to be of a novel, superior type to survive. It had to be superior in the competition of species in order to pass through the fine meshes of the evolutionary process. Only then could it produce offspring and thus ensure the survival of itself as a new species. The first difference is this: we make the selection now, not nature, and we decide what shall survive. If we do happen to make pathogenic organisms from recombinant DNA techniques, they may be submerged, if they get loose, by ecological pressures from other species. But they may not, and we thus should take great care that any species we do design are either so weakened ecologically that they will be suppressed immediately, or so designed that they could survive only in very restricted environments, with precise limits of pH or temperature, et cetera, so that life outside the laboratory laminar-flow hood would be impossible.

A second distinction is that we can do the selection much faster, so we may be—in fact, we are—in a situation where what we produce confronts us with moral choices for which we are not yet prepared. This again will be nothing new in history. Science has consistently faced us with changing moral dilemmas, but never at such speed or with such intensity.

The third distinction is that contemporary genetic engineering has relieved us from the necessity of sex. In this Brave New World it can all be done in test tubes if we so desire (an inappropriate verb under the circumstances!), and in a totalitarian dictatorship it could be done even if we don't desire. Also we now have access to a whole range of genetic variability that has hitherto been unavailable. Sexuality in nature limited the amount of genetic exchange by permitting only certain crosses. It provided just what the successful evolution of species could bear, and no more. On occasion, a plant of one species might fertilize a plant of another, or exceptionally, a plant of another genus, and the offspring might be both viable and fertile. Very rarely, animal crosses between species might produce viable but not fertile offspring, as in the case of a donkey and horse. But now the horizons and the varieties are infinite. We can make hybrids between anything we choose: between plants and bacteria; plants and animals; between totally different animals; a crocodile cell can fuse with that of a hummingbird, a spiny anteater can merge with a hippopotamus, the duck with the orange. How will these organisms behave in our world?

Our vision tells that soon we may be able to choose any desirable trait and insert it, delete it, incorporate it, amplify it, in any organisms, *if that is what we want to do,* if we want to take the time and money to travel to Alpha Centauri. Similarly, we might choose any environment in the whole world, from Death Valley to Antarctica, and engineer a plant to live there, to optimize that environment and exploit it. Other possibilities have a real streak of the macabre, if not the downright malevolent. It may be scientifically intriguing to grow a human baby in a test tube; it may be equally so to fuse a human and gorilla cell and try to grow the resulting hybrid. Indeed, it may be very easy to justify this, with a cold logic or blithe rationality, arguing, for instance, that by growing babies in test tubes, we are relieving women from the suffering and constraints of pregnancy. But while these possibilities are wonderful to some, they are downright frightening to others. I am one of those people who view them with a degree of suspicion. Peter Carlson is another, and in Part II our reasons will be developed. Our feelings have to do both with humanity, and with the nature of science, as we have traditionally understood them.

What this whole range of concerns focuses on is the question of just what it is to be human, and what kind of humanity is it we want to stand for. If all the technologies of this Brave New World are both possible—which they may be— and applied—which they may not be—then what kind of a man will we have created, and is this the kind of man we want? Scientists will rightly argue that the driving urge for more knowledge and applied knowledge is itself an important measure of man that should not be stifled. This is indeed true, yet others will insist that everything one does should be seen in terms of a wider humanity, and that, too, is what it means to be human, even if this humanness entails some pain and suffering. Qualities of compassion and responsibility, and perhaps a little "laissez-faire"—certainly a concern for the autonomy of people—must be allowed equal weight with other drives, such as logic, objectivity and the disinterested search for knowledge.

We are concerned here with the rights of individuals and the sanctity of human life, not interpreted in the strict biblical sense but with a gentle reverence. Seen in these terms, life is something that one does not manipulate lightly, *in any way whatever,* without respect or concern. It is fears like these that worry so many other people about the headlong rush into a new biomedical technology; the fear that we are losing our respect for each other, and for what it means to be human.

Admittedly it is a matter of degree. Agronomic interests have a well-deserved tradition, and current manipulations in plant biology are easy to justify. Because of its long history and its understandable aims, genetic engineering in this sense is already part of our world view. But what about these fusions? What really are the aims and motives behind those who do them? Even if the search for understanding is still a prime motive, is this enough now in a world where the advance of science impinges so deeply on the autonomy of human beings? It is an end that has the most honorable tradition in science and that has rightly been regarded as one of man's truly civilizing traits. But as we shall see, there are even some scientists who now say, "No, that end is not sufficient justification by itself for each and every experiment."

If one believes, as I do, that there has from the very beginning been a strong moral component in the practice of science which it is essential to retain for the health of the enterprise, then one must ask: "What is the moral justification for many of the experiments that are being made possible?" To mate a gorilla with a man may be fascinating, but what is the moral justification for this? Or for growing a baby in a test tube? What is the scientific question to be faced, the problem to be answered? What now is the aim of science, and can scientists justify the thrust of their work, even to themselves, given society's problems and priorities?

These are some of the questions that in the last ten years have surfaced rapidly, and which in 1975–76 I found swirling around the issue of recombinant DNA. They are very old questions, but it seems to me that the current state between science and society is like that of a supersaturated solution. A crystal of contention, recombinant DNA, has been dropped into the solution, and then as happens in a supersaturated solution, crystallization has occurred, and a whole conglomeration of issues, concerns, arguments and debates have rapidly appeared. These issues are here to stay; they will not vanish overnight; the politicians and the critics of science will see to that. In one sense it is hard luck for the recombinant DNA scientists; it was they whose work was most relevant and prominent just at the point when the crystallization was bound to occur. And they will take no comfort from the fact that their own very laudable efforts to anticipate and ward off some of the worst consequences of this technology should itself have been the very act that caused the whole issue to become suddenly and startlingly visible.

To some outsiders, and to some scientists, too, science itself, instead of simply giving us new basic data or a new understanding, seems to be transforming gradually into a techology that threatens our traditional values. Worse: people have been incorporated into that technology perceived not as people any more but as a means to someone else's ends, and science has never done this before. It seems that a total extension of cold logic and objective rationality into *anything* we do is in danger of removing the humaneness from our lives. We have, in any case, been living through a period in which we have minimized the validity and appeal of emotion, aesthetics and feelings. For I did find that, though many scientists were prepared to appreciate concerns about the effects of their work, both in the short and the long term, they were rarely prepared to weigh feelings, emotions, aesthetics or any spiritual considerations in the balance with sheer rational arguments.

Are they right not to do so? I don't know. All I know is that the traditional conflict between science and the humanities has often turned on such values, and that the reactions of many in the antiscientific movements are due to such exclusions. In any case, for a lot of other reasons, society has now decided that the time has come to rewrite the social contract between itself and this profession. This, too, has affected the course of the debates about recombinant DNA and genetic engineering. For in the end it may well be the social ethos, not the scientist's desires, that will finally determine what shall or shall not be done. It is this social ethos that we shall now examine.

II
THE
SCIENTISTS
AND
SOCIETY

5
Creating New Issues

After such knowledge, what forgiveness? Think now.
History has many cunning passages, contrived corridors,
And issues . . .

—T. S. Eliot
"Gerontion"

In 1852, General Sabine, president of the British Association for the Advancement of Science, in giving the traditional opening address said, "Scientific men cannot too highly value and desire the advantages they possess in undisturbed enjoyment of their own pursuits, untroubled by the excitements and distractions of political life." In 1975, Senator Edward Kennedy, chairman of the U.S. Senate Health Subcommittee, addressing the Harvard School of Public Health, said, "Public support which implies confidence and trust has become so substantial that many of them [scientists] believe that it is theirs by right. It is not. The plain truth is that the National Institutes of Health has been a 'sacred cow.' Those days are gone."

These two quotations illustrate better than anything else the rude awakening that is accompanying the dawn of recombinant DNA research. However, if to the outsider the image is one of a reluctant profession being dragged, kicking and screaming, into the last quarter of the twentieth century, defending its cherished privileges as best it may in the face of new external imperatives, this picture is more than a little distorted. The historical situation in which the scientific profession emerged during the nineteenth century left scientists in an extremely vulnerable position with regard to the new ethical questions of the twentieth, and so it is not surprising that they feel bewildered, even outraged, at the encroachments on their traditional freedoms.

The crucial issues are those of accountability and allegiance: to whom are scientists accountable, and to what is their ultimate allegiance? These questions have exploded in the faces of startled scientists only in the last decade or so. From a situation in which they were accountable only to themselves, and for one thing only, namely, the scrupulously honest reporting of their results, they are now being called to account not only for the money that is spent on their research, but also for the use to which their discoveries are put. Moreover, they are being asked to adopt a new attitude of responsibility towards the needs of the society that supports them, and to recognize that these should determine the goals of their research.

The scientific profession has a relationship with society quite unlike that of any of the other professions. All professions have a number of identifying characteristics, many of which the scientific profession shares. The practitioners engage in their work full-time, in a manner long accepted by their particular tradition. Admission into the profession is conditional upon a demonstrated mastery of the accepted skills and competences, to which an apprentice binds himself to conform. Professions are self-governing: they alone determine what will be the criteria for recognized accomplishment and the hierarchies of authority and eminence. Indeed, they establish the very procedures whereby authority is recognized and eminence bestowed. If a person decides to enter the profession for the status or the cash that may come his way, this is his own business, but the profession itself was not constituted for those ends. Professional activity is not only concerned with individual rewards but with something transcendental too: whether it is improved health care or a more realistic interpretation of the law or better education for the young or, as in the case of the scientific profession, the discovery of truth. With the sole exception of the scientific profession, one other thing unites them all: services to the public are their chief object, cemented by a contractual relationship. Because of this contract the competence of a professional is appraised both internally, by his contemporaries, and externally, by the public.

Accountability, in the last analysis, is the price the professions pay for their privileges. The ethnical relationship that exists between society and the professions, which marks the nature of their social contract, is supported by the twin pillars of mutual responsibility and accountability to such an extent that a series of protections has grown up between both sides, designed to shield each other from malpractice or exploitation.

Such an external accountability is missing from the social contract between the scientific profession and society, however, and this is the focus of the present dilemma. For scientists see no reason why this social contract should change, whereas society now sees every reason why it should. Scientists have held, and continue to hold, that there is no real external imperative upon them, no ethical core other than the ethics internal to the discipline. Thus, they insist, no value judgments should arise other than those necessary to determine the soundness of a scientific theory; detached neutrality and total freedom are vital for the vigorous health of their enterprise.

If this is true, was it always true? Is this privileged isolation a necessary condition for scientific growth, and indeed, is this how the founding fathers of the scientific profession meant it to be? History provides some surprising answers to these questions.

This is not the first time there has been conflict about the relationship of science to society. There was a time when the question "Who is a scientist?" was meaningless, not only because science was not a recognized professional activity, but because the very word itself had not been coined. During the seventeenth and eighteenth centuries—indeed, during the first few years of the nineteenth century—those natural philosophers, the people whose activity we recognize as scientific, were usually clergymen, like Stephen Hales, or tax collectors, like Antoine Lavoisier. It was an amateur pursuit, open to anyone who could support himself, and engaged in with varying degrees of seriousness or light-heartedness. There is a charming vignette in the diary of Colonel Roderick Murchison, who in 1823 went away for one of those famous weekends in an English country house:

> In the Summer following the hunting season 1822/3, when revisiting my old friend Morrit of Rokeby, I fell in with Sir Humphrey Davy, and experienced much gratification in his lively illustrations of great physical truths. As we shot partridges in the morning, I perceived that a man might pursue philosophy without abandoning field sports and Davy ... encouraged me to come to London and set to at science.

One may smile at the priorities of both Murchison and Sir Humphrey Davy, a distinguished physicist and well-known member of London society. The single-mindedness so characteristic of twentieth-century science was completely absent in the nineteenth century; few people took scientists seriously.

In 1830, Charles Babbage wrote a book whose effect was to be crucial: *Reflections on the Decline of Science in England and Some of Its Causes.* Babbage had been stimulated to write it by a year's traveling on the Continent in 1827, following a severe illness. There, by contrast with England, he found natural philosophers were held in great esteem, especially in France and Prussia. They sometimes even reached the highest ranks of government!

Though Babbage wrote of the "decline" of science in England, it was a moot point as to whether science there had ever risen to a point from which it could fall. Whether measured by salary or status, by honorary distinction or government post—indeed, by any criteria—science was, he wrote, in a very inferior position compared with other professions— medicine, law, the military and the church. It was not, he rightly concluded, a profession at all.

The book had the very effect that Babbage desired—a storm of debate followed by practical action, and in 1831, the first meeting of the most significant scientific society in England since the Royal Society was held in York. The British Association for the Advancement of Science was formed specifically to improve the status of science and to explain its aims. It would do the first by giving society a practical justification for the existence of science, and incidentally by strengthening the ties between the practitioners and their public in every possible way.

While the aspirations of the practitioners of science were by no means entirely altruistic, they genuinely did believe that the social contract between them and society should take the form originally defined by Rousseau: that of a relationship of mutual support and benefit. Nevertheless, the founders of the association, who referred to themselves as "cultivators of science," were also thoroughly realistic. Given the pragmatic nature of early nineteenth century English society, when scientists set out to "sell" science to society, it is not surprising that in making their case for status, they appealed not to the disinterested search for truth, but to the practical benefits that would follow from the encouragement of scientific endeavors. After visiting England, the famous German chemist Baron Justus von Liebig wrote to Michael Faraday, the British chemist: "Only those works which have a practical tendency awake attention and command respect, while the purely scientific works, which possess far greater merits, are almost unknown.... In Germany it is quite the contrary."

The efforts of the British Association for the Advancement of Science were immediately unsuccessful, however; the manufacturers were skeptical and the public indifferent, and in its early days the association was mauled, even derided. John Keble, the poet and High Anglican divine, called it, "a hodgepodge of philosophers," and in 1836–37, Charles Dickens parodied its proceedings in the *Public Life of Mr. Tulrumble, Once Mayor of Mudfog,* with full reports of the first and second meetings of the Mudfog Association for the Advancement of Everything.

Mudfog is a kind of nineteenth-century repetition of Jonathan Swift's earlier satire on the Royal Society in *Gulliver's Travels* (specifically in "The Voyage to Laputa"), though nowhere near as good. Dickens played around with his professors, Snore, Doze and Wheezy, as well as with Mr. Slug, who was celebrated for his statistical researches, and Mr. Woodensconce, who came to the meeting and gave lectures, experiments and researches in zoology, botany, anatomy, medicine, statistics and mechanical science, "plus supplementary umbugology and all ditchwater characteristics."

The members of the association were sitting ducks for parody. They met each year in a different city, as they still do, and as do their American counterparts, the American Association for the Advancement of Science. As time went on and the membership increased, however, the publicity before and during the meetings became more flowery, the hospitality more lavish and the dinner menus more lengthy. These were also published along with the scientific papers! Dickens' First Report of the Mudfog Association for the Advancement of Everything revealed that the professors and members had "balls and soirées, and suppers and great mutual complimentations." After a week of feasting and feeding, Professor Woodensconce, who had spent his time feasting and philosophizing, entered the hall "accompanied by the whole body of wonderful men," where a sumptuous feast was waiting. "Ah," says the professor, "this is what we meet for; this is what inspires us; this is what keeps us together and beckons us on; this the *spread* of science, and a glorious spread it is."

On August 18, 1838, the *Literary Gazette* took up this gastronomic theme and reprinted a report from the *Newcastle Journal,* listing the amount of game donated to the feast by several aristocratic lords, to "prove that gastronomy beats astronomy." The *Times,* reporting that same meeting, spoke of a grand promenade of some four thousand people, at which the "amusements as well as the refreshments were of the most recherché description"! As Chaudhry wrote, the general feeling was that far from popularizing science, the association had only succeeded in vulgarizing it, and these annual jaunts provided an indiscriminate mixture of science, technology, pomposity, vanity and sheer crankiness.

Even ten years after the founding of the association, the *Times* was still being impolite with the glorious direct rudeness that has characterized that illustrious newspaper from time to time. On August 21, 1841, it described a convention of some seven hundred nonconformist ministers, summoned by the Anti-Corn Law League, as a "freak" and a "drollery" no less absurd than the "British Association for the Advancement of Science." "Social humiliation" is perhaps far too strong a phrase, but in those early years it was hard going, and this may explain why the association turned to other consolations. Even then they managed to offend Victorian susceptibilities, for when they held their annual meeting at Castle Howard, near Carlisle, Lady Mary Howard, the daughter of the ninth earl, and admittedly a Victorian Carrie Nation, was so appalled by their junketings that she caused the family's wine cellar to be drowned in the lake. She even took an active part in the massacre herself by seeing to it that the top of every bottle was knocked off before it was drowned!

Why Victorian society couldn't, or wouldn't, respond to the efforts of this emerging profession to forge close links with it is an intriguing and highly complex question. The post-Napoleonic era in Europe was a period of rapid social change, and the form of the social contract that eventually emerged contained many diverse elements. We can point to one or two.

Certainly, the smug, straitlaced self-satisfaction of the newly emerging commercial classes, when added to the same attitude already existing amongst the political aristocracy, bred its own brand of anti-intellectualism, as intense as any later found in America. In addition, the Industrial Revolution was already well and truly underway, and most successfully so. The application of machine techniques to the process of manufacturing goods and to trades, crafts and arts had already produced a dramatic change in the industrial system of the country and laid the base for the great commercial wealth of the century. If the manufacturers did not immediately rush to embrace the "cultivators of science," it could be in part because they were already totally occupied with industrial applications of a science from an earlier period. In any case, the practical applications of esoteric mathematics, nerve conduction, chemical laws and chemical reactions were not immediately obvious. They never have been.

The situation has remained a feature of British life: science and industry have never really managed to forge close ties. This is doubly paradoxical so far as the early nineteenth century is concerned, for the exponents of science in England were—or so their European colleagues complained—far too occupied with the practical outcome of their work! And yet their record was a distinguished one. Plenty of foreigners gave glowing testimonials to the preceding fifty years of British science. Georges Cuvier, the famous French naturalist-paleontologist, spoke enthusiastically of its range of contributions: from explorations to astronomy; fundamental chemical laws to elementary chemical gases; the facts of heat, magnetism and electricity to comprehensive mathematical equations that embraced all these facts.

Besides genius so much else, however, was lacking: backing, money, organization, and specialist journals amongst other things. Those of the latter that existed might, as Professor Moll of Utrecht reported, be eagerly seized and avidly read by journalists as well as scientists, but more often than not scientific articles were buried in literary journals, and also in journals that could hardly be called literary. According to the *Edinburgh Review,* much good mathematics was entombed in *"The Ladies' Diary,* among poetry of the worst taste and childish scraps of literature and philosophy." That might be good for the dissemination of knowledge, but it was lethal for the professionalization of science!

Money too was sadly lacking. Whereas French chemists like Joseph Gay-Lussac and Henri Regnault were backed by wealthy academies, Michael Faraday actually supported the Royal Institution that employed him, keeping it alive financially, by the lectures *he* gave! "We were living," he once said to its managers, "on the parings of our own skin." It was an uphill struggle all the way.

By the 1840's, though there were sufficient practitioners for the word "scientist" to have been coined by William Whewell, the Master of Trinity College, Cambridge, little progress had been made. Scientists were still regarded as a comparatively low status group, and the rationale with which the British association, in its first report, pleaded their cause, still held. "There is a service of science to be rendered to the State with which it cannot dispense. . . . Therefore, all, I think, must allow that it is neither liberal nor politic to keep those, who employ the rarest intellectual endowments in the direct service of the country, upon a kind of *parish allowance.*" It was a long time indeed before science provided a secure living.

Eighteen fifty-one was the year of that magnificent display of British enterprise and commerce, the Great Exhibition, and by that time too, key changes could be observed in the relationship between the scientific profession and English society. That shift is quite unmistakable in the presidential addresses of the first eleven years of the association (1831–1842) compared to those of the next fourteen (1842–1856). The focus and emphasis increasingly turns inwards toward the methods of science: intellectual understanding becomes the aim of the profession, and truth its only goal.

Part of this change was undoubtedly due to reaction; to an extent the scientists' earlier experiences prompted a quite deliberate withdrawal from society. There was more to it than that, however. The Romantic German attitude towards learning had been imported into England through many channels, one of which was Prince Albert, Queen Victoria's consort, and had come to predominate in science. According to this philosophy, the pursuit of knowledge for its own sake was the highest goal to which man's spirit might aspire, and those who, in the hallowed halls of learning—which meant the universities—devoted their lives to this end represented a pinnacle of man's achievement. Second, as time went on, the tendency towards autonomy and introversion became an almost natural extension of science's becoming more professional; since there were no services to be performed to the public, which in other professions were a precondition for existence, this tendency intensified. If society didn't really need the scientific profession, it is equally true to say that in those early years the scientific profession didn't need society either.

By 1855, when the Duke of Argyle gave the presidential address, the concern of the community, its very *raison d'être,* had become the disinterested search for truth, a search undertaken for its own ends with little regard for the practical benefits which might follow. Following the great university reform movements of the 1850's through 1870's, which greatly improved standards of teaching and attainment, the seal was set. Scientists and their research were firmly accom-

modated within the framework of the universities. By the
last quarter of the century, the direct relationship of the
profession to society was tenuous to say the least, and the
British Association was paying only lip service to the practical application of science. The aim of the community was to
preserve those conditions which seemed to them to be most
compatible with the advance of scientific knowledge: isolation and autonomy. These conditions had never been proven
to be prerequisites for the advance of knowledge, but like
Topsy, the situation "just growed." Science still remained an
insecure profession, though, and Thomas Henry Huxley
could still complain that no amount of proficiency in the
biological sciences "will surely be convertible into bread and
cheese."

These nineteenth-century attitudes had several important
implications for the practice of science in the twentieth. The
most significant of these, so far as recombinant DNA research is concerned, relates to the questions of allegiance and
accountability. In 1852, the year in which General Sabine
reminded the profession of their singular advantages, Sir
Lyon Playfair, lecturing retrospectively on the results of the
Great Exhibition (*The Chemical Principles Involved in the
Manfactures of the Exhibition ...*), posed the question directly: to what does science owe its ultimate allegiance? His
answer was "To the sublime truth." Prince Albert, who was
for sublimity in all things, agreed with him, and on the
whole, scientists were most willing to accept what the
Prince Consort described as a "self-conscious abnegation for
the purpose of protecting the purity and simplicity of their
sacred task." Romanticism can go no deeper!

But if the ultimate allegiance was to truth, the next step
followed logically: this allegiance was not to society. In particular, abstention from any form of politics was not only
desirable but a necessary condition for valid scientific study.
The job of a scientist was to supply the facts and stand aside.
Taking a political stance on anything was tantamount to
distorting the results of one's scientific investigations. Four
years earlier, in 1848, an engineer named Seymour was bitterly attacked for agreeing to sit on Edwin Chadwick's Social
Commission on Sanitary Reform, and there was, indeed, a
move to deprive him of British association membership.

This issue of politics and science would become a hot one in the twentieth century.

The situation in American was no different from that in England except in degree. In a country which, even as early as the nineteenth century, had a fantastic record for applied technology, recognized as of world importance, the neglect of basic science was a puzzle. Once again the social situation was complex enough for the causes of such attitudes. Certainly America had enough centers of population and wealth to support basic research inside or outside their universities. They had the other conditions too: a tradition of free inquiry, a faith in the belief and hope of progress, no fossilized prejudices or attitudes, no persecution for those who challenged cherished ideas, a strong and distinctive literature. De Tocqueville, a wonderfully perceptive observer of the American scene, commented on the American indifference to basic science. He pointed out that they in fact could borrow everything they wanted from Europe, but they didn't. He attributed this indifference in part to the combined factors of democracy and economic opportunity. To a degree his observations are still valid. In an egalitarian society, he said, the mass of people would support science only to the extent that it was a tool for exploiting natural wealth for *their* own individual material interests. He wrote: "In aristocratic ages, science is more particularly called upon to furnish gratification to the mind; in democracies, to the body," and in America there were no leisured or rich classes, the groups that in Europe were traditionally free to gratify their minds with science.

Given the totally different societies of England and America in the nineteenth century, it is intriguing to see both countries using the same strict utilitarian yardstick with which to measure the value of science. America applied the same measure in the universities as well. Anyone who chose to pursue science there found that he was valued as a teacher only, and not too highly valued at that! The highest praise and esteem was given to the practical or "useful" person.

It was quite different in Continental Europe. Science was highly respected in France and Germany—and in France there was even a good amount of interaction between theoretical knowledge and technology. The history of the profession in Germany, however, shows again how it became isolated. Scientists were uplifted, instead of downgraded, but the effect was the same. Swollen with their own importance, they became, as Fritz Langer has called them, "the mandarins." They felt it was society's job to serve *them* by ensuring that the conditions for carrying on their sublime task were maintained. The profession's attitude became, and remained, thoroughly aristocratic! The famous explorer-naturalist Alexander von Humboldt expressed a sentiment which sounds very familiar by now: "Science is the fundamental thing, for when it is pure, it will be correctly and sincerely pursued, in spite of exceptional aberrations. Solitude and freedom are the principles prevailing in its realm."

The imprimatur given by Prince Albert and Baron von Humboldt may have been typical pieces of Victorian Romanticism, but they, too, evoke nostalgia in the breasts of a number of contemporary scientists, still willing to accept "self-conscious abnegation for the purpose of protecting the purity and simplicity of their sacred task."

These attitudes lingered into the twentieth century. It took many decades before England, let alone America, fully accepted the work and value of scientific studies, and the profession did not reach a point of status and public support comparable to that in Continental Europe until after World War II, Einstein and Sputnik. Between the two wars, American scientists flocked to Germany: J. Robert Oppenheimer found his spiritual home at Göttingen, as did many other scientists. It was a natural place to gravitate because of the sheer intellectual excitement—the brilliant men there and in Cambridge, England, had begun to open up the new world of physics—and also because for those scientists with detached monastic tendencies, the apolitical backwaters of Europe were intellectual havens. They were, of course, highly privileged to live in the world of mind alone, but the price was to be paid very soon.

Oppenheimer, who could well serve as our touchstone, found that his experiences in Germany in no way prepared him for the political problems that he was to face during the 1950's. Then the issues of civilian versus military control of the newly created Atomic Energy Commission, and dissemination of scientific knowledge versus increased secrecy, brought to a head conflicts that had been festering ever since the successful completion of the Manhattan Project. That episode notwithstanding, the apolitical ivory-tower tradition became the accepted one, for better or worse. This was the inheritance that European and American scientists came to take for granted, so much so that any other form of social contract could hardly be conceived.

This relationship was not so much due to deliberate acts of professional aggrandizement on the part of the practitioners of science—though there were these, of course—as it was to the fact that as the profession emerged it was swept along by the tides of prevailing social thought. Equally, society got the profession it encouraged, as in Germany, or the one it "ignored," as in England. From a combination of indifference and a demand for utility—a demand which still haunts science today—Western society found that it had fostered a highly privileged—some say societally amoral—group of independent people whose work suddenly came to have a most startling power.

The end result was a profession quite unlike any other. It had acquired a high degree of privilege, and by default, an equally high degree of public support. This, in America, was a direct result of the post-Sputnik investment in science, when the future existence of the country came to be seen in terms of a strong science and technology. Save for the expenditure of society's funds, however, the profession was still accountable for nothing. The law in no way held them to the highest degree of care: they were never sued for malpractice nor for misapplication of their work. The only set of ethical principles that ever concerned them were those concerned with protecting the good name of the profession and its "sublime" methodology. They were in no way concerned with the needs of society; until quite recently everyone was happy with the status quo.

But not any longer. It is certainly hard luck for the scientific profession that after all these years with society apparently being quite content to allow it all its professional privileges, it is suddenly being required to settle the accounts. But history always exacts its price, and the profession has already begun to pay. One hundred and twenty years after General Sabine gave his presidential address, the British association met again in Stirling, Scotland. The president for 1972 was Sir John Kendrew, a molecular biologist, and as usual, it was incumbent on him to deliver a presidential address. Like those of his distinguished predecessors, it reflected something of the problems facing scientists and the present ambience in which they were working, as the president perceived them. Kendrew argued "that the concept of knowledge as absolute good in itself, which was the prime motive of scientific research, [needs] to be re-examined. It [can] not be the only motive . . . it [is] neither socially productive, nor intellectually satisfying, to refuse to modify old attitudes . . . or to respond to new challenges."

If this challenge produces conflicting responses amongst scientists, especially among those young Turks who are working on recombinant DNA, it evokes sympathetic echoes in many nonscientific minds. Once again, recombinant DNA is the pivot around which the controversies turn. While scientists push for extensive programs, public reaction has already halted the momentum, ironically and unpredictably as a result of the scientists' own initiatives. It is to these that we must now turn.

6
Creating New Commandments

Reading maketh a full man; conference a ready man; and writing an exact man.

—Francis Bacon
"Of Studies," *Essays 50*

In *The Alexandria Quartet,* Lawrence Durrell wrote about the same set of circumstances as seen through the eyes of four individuals. All four books were linked by one theme: how in details and interpretation a given situation can be perceived in vastly different ways by people caught up within it. The same is true of the Asilomar Conference. In February 1975, at Asilomar on the Monterey Penisula in California, nearly one hundred and forty scientists from all over the world met. There, diverted neither by the blue sea nor by the migrating Monarch butterflies which each spring envelop the fields and trees in immense curtains of orange and black, they tried to thrash out some new commandments to govern the directions of recombinant molecular research and regulate the procedures that could control safety. Talking afterwards to many of the people who attended, listening to the tape recordings of the proceedings (which on application can be heard in the offices of the National Academy of Sciences), and reading what was written about the conference—from the extremely formal report submitted to the U.S. National Academy of Sciences, summarized in *Science* (July 6, 1975), to the light and enchanting piece written by Michael Rogers in *Rolling Stone,* June 19, 1975—I came to wonder whether we were all speaking about the same occasion. It struck everyone so differently, though the object was defined clearly enough: "Here we are," a young scientist told Michael Rogers, ". . . sitting in a chapel, next to the ocean, huddled around a forbidden tree, trying to create some new Commandments—and there's no goddamn Moses in sight!"

I suspect that the need for Moses did not become apparent until the end of the conference. Certainly two years earlier, when the first stirrings of that concern and activity that would climax at the Asilomar Conference began, had you asked any scientists to prophesy the outcome, they would probably have said no more than "Well, we'll try to agree on a few safety procedures." What happened at this conference, and most especially during its aftermath, however, left scientists holding a Pandora's box of their very own. They had opened one for us, containing the promise of hazardous bugs: they now found they had opened a second box as well, equally dangerous for them—or at least so some scientists have felt. It contained such hidden perils as public participation, intervention and interference. As a result of Asilomar, onto the seemingly gentle world of scientific research has descended not gentle migrating butterflies, but hordes of public and political gadflies.

Private concerns about recombinant DNA had first been expressed to Dr. Paul Berg as early as 1971 by a young cancer researcher, Robert Pollack, but the public story really begins in 1973. In January of that year a small meeting was held at the Massachusetts Institute of Technology, a meeting that scientists now refer to as Asilomar 1. In the words of Berg, it revealed "a recognition of how little we know." Active moves began later that year, and Dr. Maxine Singer, head of the Nucleic Acid Enzymology Section in the Laboratory of Biochemistry at the National Cancer Institute, and one of the key members of the Asilomar organizing committee, gave me the details. Each year a conference on nucleic acids, known as the Gordon Conference, is held in the United States. Maxine Singer was cochairman at the meeting in 1973 during which the joining experiments were first described. It was a very dramatic occasion. People quickly appreciated just how beautiful a tool the procedure would be, and listened, fascinated. "My goodness," one said, "you can join any old thing you want." It was an ecstatic prospect to some, but others quickly reacted with concern. The ethical and moral problems involved in decisions to engineer desirable traits in complex organisms, or to conduct experiments which might threaten the safety of others, were immediately apparent to a number of scientists—and in those socially activist times, they could not be ignored.

By the end of the conference several people had approached Maxine Singer and her cochairman, Dieter Soll of Yale University, to express their reservations, and after some discussion, they decided to throw fifteen minutes of the last morning's session open for comment. Meanwhile, they urged the scientists to talk about it during "bar time" in the late evening hours. The next day, realizing how short their time was, those assembled voted, with little disagreement, to send a letter to the National Academy of Sciences pointing out that a problem had been raised which should be investigated.

There was quite a bit more disagreement about the next suggestion, however, which was to "go public" and send the letter to *Science* as well. Many of the scientists were already very wary of public involvement at this—indeed, at any— stage. In their minds, "going public" was, and still is, an invitation for an antiscience reaction: some things were best kept quiet. But on the wishes of the majority, the letter was published in *Science* on December 21, 1973.

For the first time, the possibilities—and possible hazards —of recombinant DNA were raised publicly. The reaction of the national academy was swift. Even before the letter's publication in *Science,* Maxine Singer was asked to Washington to suggest who should serve on a committee to examine the problem. Her first and most natural choice was Dr. Paul Berg. Not only was Berg critically aware of the implications of the work, since most of the techniques had been developed at Stanford, but he was a man of enormous integrity who had a great deal of stature in the biological community. It was already obvious that whoever undertook this delicate and diplomatic task would need to command a fair amount of respect from his contemporaries.

The history of the committee was already unique. As Dr. Singer recalls, the scientists' rapid recognition of problems in an area of work that had only just started *was* rare, and she knows of no other instance where an informal group has gone to a larger scientific institution like the academy and specifically asked it to take over a problem. What happened then, however, was even more unusual. Paul Berg quickly chose eleven distinguished and established scientists, such as James Watson; Norton Zinder of the Rockefeller University; Stanley Cohen (of PSC 101 fame); David Baltimore, who was later to receive the Nobel Prize for his work on RNA tumor viruses; and Daniel Nathans and Sherman Weisman, who had been at the earlier Gordon Conference. The result of their discussions was the "Berg Letter," which was published in the July 24, 1974, issue of *Science.*

It was something of a bombshell, though it was very explicit. It recommended that until the potential hazards of recombinant DNA molecules had been properly evaluated, or until adequate methods for preventing their spread had been developed, scientists throughout the world should voluntarily defer three types of experiments: constructing replicating plasmids that would introduce either antibiotic resistance or bacterial poisons into bacterial strains, and linking DNA from likely cancer-causing viruses to bacterial plasmids. Moreover, they should think carefully about the pros and cons of whether fragments of DNA should be linked *at all* at this stage, given the fact that many sequences of animal-cell DNA were known to be common to RNA tumor viruses, and that the biological properties of the new recombinant DNA molecules could not be predicted with certainty. In addition, they suggested that the director of the National Institutes of Health establish an advisory committee to develop an experimental program to evaluate the hazards, develop procedures which would minimize the spread of such molecules within populations, and devise guidelines to be followed by investigators. Lastly, they called for an international meeting of scientists to review scientific progress and to discuss appropriate ways of dealing with a potential hazard.

Maxine Singer remembers being somewhat surprised at the breadth of the recommendations, but also felt the letter reflected a very sensible approach to the problem. What she didn't foresee—in fact what none of them foresaw—was the extent to which all of their lives were to be dominated by the consequences for a long time to come—especially since no one on the committee had ever thought of how, having got yourself into a moratorium, you got yourself out! The committee's report was only an "ad hoc" response—though a thoroughly responsible one—to a danger which had been mooted in a small way by a group of people with very limited experience in public affairs—but by now they were deeply enmeshed in something very public.

It is important also to remember that *everyone* concerned felt that they had really called for a "pause," not a "moratorium"—and they now rather wish that this had been the word used. To suppose that a semantic change could have altered the situation is a little naive, but the scientists didn't like the dramatic sound of the word "moratorium"—and especially how this would be received by the media. Moreover, the notion of a "pause" had a continuity about it—and this is the real point, for all the scientists expected to go ahead and do the experiments in the end.

The moratorium was the most dramatic aspect of the Berg Letter, but the other recommendations were equally important. The Gordon Conference had gone to the academy, not to a government body, because they viewed the academy as the spokesman for the scientific community. The Berg committee took a different approach, however, recognizing implicitly that this was a *public* issue, one which rightly belonged to an agency with official and public standing. Shifting the onus to the National Institutes of Health was a minor recommendation in the Berg Letter, but a very telling one, because it reflected that at least some members of the committee already recognized the governmental and international aspects of the situation—that certain issues very fundamental to the relationship of the scientific community to society were at stake.

Science is fearsomely competitive, and it is not surprising that some individuals bitterly resented the moratorium, privately believing that it had been called only so certain groups, especially the one at Stanford, could maintain their advantages in recombinant DNA research. Charges and countercharges multiplied, most of them unfair, however. Though—as was obvious at Asilomar—a number of scientists had done work that was seemingly proscribed by the moratorium, the Berg Letter, had actually said, with some ambiguity, that "such experiments should not be undertaken lightly." Thus it clearly became a matter of personal judgment whether or not to proceed. But *no* scientist disobeyed the injunction to hold off on the three specified types of experiment.

In the midst of all this confusion, the international conference called for by the Berg Letter was being organized by Maxine Singer, Sydney Brenner, Paul Berg, David Baltimore and Richard Roblin, through meetings, correspondence and one enormous and expensive conference call which linked Washington, Stanford, Boston and Cambridge, England. It was a hectic time. When I commented to Dr. Singer that it sounded like having another baby arrive in the house unexpectedly, she replied tersely: "Not half as much fun."

Their first problem was to select the participants. Molecular biologists about to work or already working in the field obviously had to be invited, but other needs had to be fulfilled as well, for it was clear molecular biologists and biochemists just didn't have certain kinds of knowledge. Their ignorance about such matters as epidemiology, for instance, had been at the root of many of the committee's previous difficulties in thinking constructively about the problems. Even though very little was known about the ecology and spread of *E. coli,* such experts as there were clearly had to be present. In addition, they needed agricultural biochemists; scientific administrators responsible for funding research and making policy from NIH and the National Science Foundation; members of the Policy Committee of the European Molecular Biology Organization (EMBO); and representatives of the research arms of drug companies, such as Hoffmann–La Roche, Merck and GE.

It was also clear that people with nonscientific interests should be present. Maxine Singer's husband, Dan, was a lawyer associated with the Institute of Society, Ethics and the Life Sciences, in Hastings-on-Hudson, New York, whose members had already had a tremendous impact on the debates over ethical issues in biomedical science. He had already discussed these matters with Paul Berg several times, and so it was quite logical that he be asked to invite a group of lawyers to the conference. As it turned out, they were to make a most unusual contribution to the proceedings.

It was more or less assumed that the meeting would be open to the press, but here some constraints were imposed. If reporters came at all, the committee decided, they would have to come for the whole meeting and stay at the center as members of the conference, taking their meals with everybody. Though everything would be quite open to them, they would be asked not to file any stories until the very end, since the situation was likely to be very fluid. New issues and counterissues, fresh moods and attitudes might arise every day—as indeed they did—and it would be difficult to capture the distinctive essence any other way. Such restrictions took some swallowing; many journalists thought it unlikely that they could sell the idea to their editors. Newspaper editors are happy to send journalists off on jaunts to the sun, but only if they are confident of a story every day. The organizing committee was adamant, however, and later opinions seem to have been unanimous. The press thought it was a marvelous device, for their stories were better, and the scientists agreed that the newspaper accounts reached a level of understanding and responsibility rarely found in the daily-newspaper coverage of science.

What did the organizers think the conference would do, and what did it actually do? Dr. David Baltimore of MIT, introducing the first session, emphasized that the members present were a small, select group of scientific individuals, and that though the issue was one of safety, perhaps even morality, the focus of the meeting was to be primarily scientific. All discussions would have to be grounded in the *facts* of the situation. The peripheral issues, such as the utilization of the technology for genetic engineering and the potential for biological warfare, were to be outside the scope of the meeting. (Some of the younger scientists were most unhappy about this.) The main issue was this: the profession now faced a radically new situation, with new technologies at their disposal that would allow them to override and outdo the events of evolution. How should they proceed? What could they do in the face of uncertainty about both the hazards and the benefits?

At least four questions had to be discussed: (1) What were the benefits to be derived, so that the public would understand the "urgency" of moving into the future? (2) Should the safety procedures emphasize physical containment facilities or biological safety? (3) Should scientists do experiments that would enable them to assess the degree of future hazard and its nature? (4) How much national and international compliance should there be with these procedures, and how could this compliance be assured?

David Baltimore insisted that it was to be a scientific meeting, and it certainly was that. Nearly everyone believed that the sole reason for coming to Asilomar was to decide what had to be done to get research started again—and fast! Many of the younger scientists were already extremely impatient. As the first provisional statement of the organizing committee later put it, "The participants at the meeting agreed that the 'pause' in research called for in the July 1974 Committee letter ought not to be left unresolved." In addition, it was soon obvious that the scientists were happy to talk about *anything* rather than the societal issues at stake, and preferred to retreat into a maze of technical details at the first signs of difficulty or uncertainty. Time and time again, whenever more general issues, especially those outside science, came up, the scientists tended to submerge them with a display of experimental and technical pyrotechnics, dazzling themselves and each other. And so it began—and continued!

After an intellectual demonstration that K12, the laboratory form of *E. coli*, could indeed survive in the human gut long enough to exchange genetic material, even though it was very much weakened by the genetic manhandling of years of laboratory research, the scientists settled down to discuss the papers of three working groups. The first group, under Richard Novick of the Public Health Research Institute of New York, looked at the hazards of using plasmids or phages, and tried to classify possible experiments into six categories, according to their likely degree of danger, ranging from simple work on bacteria that were known to recombine in nature anyway to experiments on the most dangerous pathogenic bacteria. For each class, they recommended the procedures of physical containment which, in their judgment, would reduce the hazard to an acceptable level.

The difficulties of the problem emerged at once, for as Dr. Sydney Brenner remarked, "One of the biggest biohazards is our lack of knowledge." It was child's play to find areas of legitimate disagreement. Who is really to say that there is less risk in using DNA from a toad than DNA from a bacillus? Some criticized the document for being too vague; others, such as Dr. Joshua Lederberg of Stanford University, attacked it for being too precise! Lederberg feared that legislators would proceed to regard these classifications as the definitive statement from "those who know best" and so rule out any further discussion or development. The dilemma is a perpetual one. If an injunction is worded too precisely, it loses flexibility and can no longer adapt to changing circumstances, but if it is worded too vaguely, its very ambiguity deprives it of any force.

Even more important than Lederberg's objection—and probably the single most important statement made at the conference—was Sydney Brenner's reminder that the conference should in no way regard itself as a body able to grant legislative authority to eager scientists. Throughout the conference, he continued to emphasize the political and social note: the antiscientific attitude in society was already well developed, and whatever they did at Asilomar must not add fuel to that fire. Who, he asked, actually believes by now that pure science will increase the gross national product? Was there a chance that, in this particular issue, the scientific profession would be able to regulate itself? Because if they failed to do so, they would lose. Britain already had strict regulation in one field: animal experimentation. It was all very well talking vaguely about policing, but there was no way out. The profession *must* be seen to be acting. Brenner said he preferred someone who would stand up and say, "The whole thing is a lot of nonsense. I won't be regulated by anyone and I'll do what I like"—for that at least was honest —to someone who said, "We will stick the safety notice up and go on, more or less, with the guidelines, but we won't be changed at heart." Such an attitude would be fatal. He reminded everybody once again that the issue was "how to proceed in this area, without presenting any risk to ourselves, to the innocent within our institutions, or to the innocent outside them. I think there are people here who feel that there will be a *negotiable* set of compartments, and that any particular compartment must comply with their local conditions. I am utterly opposed to that way of thinking. If people think they are going to get a license from this meeting, a notice they can put upon their door—if they are just *pretending* there is a hazard and going along with it just so they can get tenure and be elected to the National Academy, and other things scientists are interested in doing, then the conference will utterly have failed."

Brenner touched a nerve as he searched out those individuals who saw themselves as accountable to no one but themselves. Later, the question of "Class 6 experiments" would come up—those experiments so dangerous they should not yet be attempted. Though the vast majority—85 to 90 percent of the scientists who voted—were in favor of such a restricted category, the remaining 10 percent voted—and always will vote—against *any form* of accountability. Their position was and still is: no one—neither public agency nor peer group nor society—can say to a scientist, "Thou shalt not do a particular experiment."

Another issue came up in the same session, initiated wittingly or unwittingly by Dr. James Watson: technical sloppiness. He blandly announced that he thought the moratorium should end, a statement which took many people greatly by surprise, since he was a signatory of the letter which had called for it in the first place. He said that, after consideration, he felt that the work in his laboratory on tumor viruses was infinitely more dangerous than anything that was likely to go on with recombinant DNA, where the dangers involved "are probably no greater than working in a hospital." He still holds this position. It is a fact that cancer virologists and old-style microbiologists have been dealing with highly infectious agents for years. At the conference these scientists felt that if the people working with *E. coli* were "screaming," it was because they were inexperienced in safe techniques for handling dangerous materials. The general carelessness of the younger molecular biologists appalled the old guard. They even described it as "prostitution" of microbiological techniques. People were just scientific slobs—"Most of my colleagues are," one scientist said to me. The cultures get poured down the sink and so into the sewers. Viruses and bacteria are treated as loosely as powder substances out of a jar or detergent out of a bottle. What was needed, in their view, was not restrictive legislative guidelines, but simply good old-fashioned training and discipline in careful microbiological techniques, together with a modicum of common sense. There were even some who took the attitude that it was solely up to the individual scientist whether he took any notice of these precautionary techniques or not. If he got a bad infection because of his own carelessness, that was his own fault. That session was lively; the next one was even more so.

The second working party had considered the problems that arose from the use of the genes of animal viruses. The document they submitted was very short and had a dissenting minority of one. It turned out, however, to be extremely controversial from the word "go." One participant said to me that the conference was just ripe for an intellectual "punch-up" at that time anyway, and the slim document lent itself well. For a start, the panel represented extremes of opinion. On one side was Dr. Bernard Roizman of the Experimental Biology Unit of the University of Chicago, who said there was no way he could ever accept limitations on a particular experiment. On the other was the minority report of Dr. Andrew M. Lewis of the Laboratory of Viral Disease at the National Institute of Allergy and Infectious Diseases, which advised "slowing down" the work. He believed that the risks associated with the widespread use of only semi-contained animal virus procedures exceeded the reward for the information to be obtained. He took this stance, he said, "because of these risks, [and] the diversity of opinion amongst individual scientists that these risks have provoked, and the moral and legal issues raised by the rising public concern about rapid advances in biomedical technology, I believe that application of this procedure to study animal virus genomes requires a methodical and carefully conceived attempt to reduce the risks to more acceptable levels." He called for the development and testing of new, safe vectors. Only when they had been provided should the recombinant studies begin, and even these should start with the nonvirulent regions of genes only. He advocated no ban, but an extremely cautious, step-by-step approach. His view was highly unpopular.

But he had a reason for his stance—his own personal "horror" story of the problem of dealing with these molecules. He and his colleagues had made a series of hybrids between the SV40 virus and the Adenovirus 2 (a virus common in man), which infects adenoids, tonsils and lymphoid tissue, especially in children. Because of the speed and ease with which humans could be infected by the adenovirus, his group concluded they couldn't possibly predict how widespread an infection would result should the hybrid spread. They had already sent one set of the hybrids to other workers, with a statement expressing concern about the hazards, listing the precautions they themselves were taking, and asking the recipients to refer all requests for the hybrid virus to Dr. Lewis's laboratory. At least one laboratory ignored the request. Dr. Lewis was quickly confronted with threats of administrative and legislative action from both sides: by those who insisted on unconditional release of the hybrid, and by those who supported its containment! Consequently, before distributing other samples, Dr. Lewis asked for a statement from his director about their release and distribution.

In December 1972, the deputy director for science at NIH, stated that containment facilities *must* be used to grow these agents. By February 1973, since he personally did not have these facilities, Dr. Lewis stopped growing the viruses. Finally there emerged a "memorandum of understanding and agreement" to cover the use and release of these viruses, which became official policy. By the end of 1973, all thirteen laboratories that had requested stocks of the virus in the first place had received not the virus but copies of the memorandum, which listed the safety precautions for containing the virus within the laboratory. Before the hybrid virus would be sent, the signature, and thereby the assent, of both the investigators and the director of the institution were required. By February 1975, only the Salk Institute returned the completed document, and therefore only the Salk Institute received seed stocks of the virus. Dr. Lewis concluded that the reasons for the reluctance of others to sign included "the belief that [it] places unnecessary restrictions on the freedom of scientific inquiry."

The right to freedom of inquiry: it is an issue that has remained in the center of the picture ever since Asilomar. Even Dr. Paul Berg was against the Class 6 category as an unnecessary and dangerous restriction on that freedom. Though Dr. Lewis had found his experiences very discouraging, after Asilomar he spoke optimistically. "I sensed a change in the meeting, and I think that a number of factors, including peer pressure have and will promote a change in the attitude on the part of some scientists."

The third working party looked at the problem of cloning DNA of higher organisms. All such experiments were either moderate or high risk. Since these would require vectors to carry in the hybrid bugs, and host cells to grow them on, the conference turned to the possibility that biological rather than physical containment might be the most effective way of handling the risks. Thus the charming concept of the "self-destruct bug" came into science—the bug which simply could not live outside the laboratory environment, and so would be automatically destroyed if it escaped. Even at the time of the conference, Dr. Roy Curtiss of the University of Alabama claimed that he had made new disabled strains of *E. coli* which could survive only in a very restricted range of environment. The call for "safe vectors" incited the scientists to strive for heights of imagination and invention as they thought up new specifications. The bugs should carry no unnecessary genetic information; they should be unable to link with the host's cells and chromosomes; they should have mutations that would prevent them from living at the same temperatures as mammalian body cells; built-in mutations should be irreversible; everything should be done to confine the plasmid to its original host bacterium—other mutations could be added which would lower the probabil-

ity of escape from the host, or a segment could be inserted
that would put the bug at an ecological disadvantage if it
escaped. Not only could the bug be made completely dependent on the existence of its host, but the host itself could be
made dependent upon special laboratory conditions: a sort of
"dual safety" feature. The challenge to science to create these
new safety bugs is enormous. Brenner, with characteristic
British understatement, called it "interesting science," but
others like Dr. Zinder, recalling that for years scientists have
handled the most dangerous microbes safely on the basis of
physical containment alone, feel that though it was the right
and honest thing to suggest, it was possibly a strategic error.

And so the scientific meeting continued, absorbed with
technical issues and internal disturbances, with only occasional forays into outside issues. But if the scientists tried to
keep the scientific blinkers on, three lawyers in the penulti-
mate session ripped them off, shattering the happy comfort
of the conference, with a few home truths about science and
the law, the relationship of science to society, and the legal
and moral rights of the public. The scalpel-like efficiency
with which they went to work and painted other different
but equally alarming scenarios astonished and startled the
scientists. (Painting scenarios was a favorite occupation at
the conference, as the lawyers pointed out.) They laid bare
a whole series of frightening possibilities, this time involving not human epidemics or pandemics but multimillion-
dollar lawsuits and injunctions against research that could
descend on scientists and the owners of laboratories whether
or not they meant to do harm. The bugs made by recombinant DNA may very easily fit into the legal category of "mischievous things," for whose escape the scientists and their
employers would be responsible and for whose subsequent
damage they would undoubtedly be liable. Not only that, but
any damage or mischief or injury would be ultimately assessed not by a panel of scientific experts nor by sympathetic
political administrators but instead by a jury—possibly a
jury made up of people antagonistic to science. This was a
new and very serious reason for the scientists both to be
cautious and to carry the public along with them. How many
scientists went home and extended their personal liability

insurance after that session would be a fascinating statistic to investigate.

The legal and political ramifications of this issue are so complex that they require a chapter to themselves (Chapter 8). For now it is enough to say that though calm over these legal problems seems to have descended, it is a careless and deceptive one. "They really were just dramatizing a bit," one of the scientists said to me hopefully. "They were not really serious, though." He is wrong; the lawyers were very serious, and yet as the memory of that session fades, the scientists seem to have recoiled from these horrifying possibilities, in the same way as one takes refuge in the thought that once an immediate danger is passed, it will never occur again. Yet as this book goes to press, the Friends of the Earth announced that they were taking the whole issue to court, to ban the use of federal funds for this research.

Maxine Singer recalled the conference for me in one succinct word: exhaustion. The organizing committee, joined by Dr. Norton Zinder, spent day and night in their sessions, and any spare time went to drafting a final document they hoped would receive the approval of the conference. Because they worked well into the small hours and were unable to take part in any of the informal conversations, Maxine Singer felt they missed the sense of things very much—not only the flavors but the general drift of feelings.

On the evening before the final session, they put their final document together with the help of a number of people who drifted in and out. The next morning they would present it to the whole conference, and they wondered how best to handle it. Certainly it should be discussed section by section, but they assumed it would be a disaster if they put it to the vote. There would be tremendous rows over the details— heat perhaps, but not necessarily light. What was needed was a strong document saying that by and large the experiments could continue, but only under certain restrictions. Their impression was that they might not have the support of the conference for that. The moratorium—"pause"—certainly *had* to end somehow, but because of the contentious nature of the issues, expressed most vocally, the conference could have dissolved into chaos, everyone going their own way, and nothing being resolved.

They had clearly underestimated both their colleagues' willingness to be put under some restraints and the moral authority evoked by Dr. Paul Berg. It is true that in all meetings of headstrong groups, it is usually those who are the most vocal and dominating who get heard, so one suspects the organizing committee made the natural mistake of equating decibels with strength of opinion.

The last session gave *them* their great surprise. Paul Berg began by saying that the document was the organizing committee's report and they wanted everyone's comments on it. Immediately there were questions as to why it was *just* the organizing committee's report? So then they decided to let the conference discuss it section by section. When they came to the end of the first section—a very simple one essentially giving the history of the "pause"—somebody said, "We ought to have a vote." Paul Berg immediately agreed, and the rest of the committee, astounded, stared hard at him, remembering their decisions of the night before. Dr. Singer still isn't sure whether Dr. Berg thought he could take a vote on that section alone or whether he had very quickly sensed the cooperative feeling of the meeting with regard to the total report. But all their worries were groundless: the vote was overwhelmingly in favor of the introduction of certain restrictive principles, already implied, and they continued to be overwhelmingly in favor of every one of the committee's recommendations. The dissensions ranged in number from three over one issue to a dozen over another. The final document was strong, neither ambiguous nor vague.

Ironically, the recommendations of the conference consolidated, even extended, those of the moratorium. The bulk of the work on recombinant DNA would continue, it said, but with certain restrictions relating to safety and containment requirements. Certain experiments were still considered to be so hazardous as to be banned: those that conferred antibiotic resistance onto bacteria, and those that involved known oncogenic (tumor-causing) viruses. In addition, there were two other recommendations not mentioned in the moratorium. First, the quantity of materials, fluids and so forth that could be used or created was not to exceed ten liters of culture fluid—a quantity ample for a scientist, but a drop in the bucket for a drug company that wanted to manufacture a human protein like insulin by this method. Large-scale work was thus proscribed. Second, the remaining groups of experiments should be ranked according to danger on the best judgment possible, given existing knowledge and information; then the physical or biological containment of the experiments should be matched to the judged danger.

A few questions still remained very controversial at the end, but that was expected. Some people thought and still think that the whole exercise was a mistake and the initial procedure wrong. No one should attempt to restrict research or scholarship in any way, they felt, and by doing so in such a public manner, the scientists invited all sorts of action which in the long run would be to the detriment of the enterprise. A young scientist told me that there was some division according to age, though not much. The older people tended to fall into two groups. There were those who were very cautious indeed about the image of the scientific profession, perhaps because they had lived through the whole of World War II and remembered what had happened with the doctors in Nazi Germany. These people were very concerned over the whole question of social responsibility. There were also people like Dr. Joshua Lederberg, who argued, very strongly and emotively, that since the research really concerned the possibility of saving lives, this was far more important than the small risks involved.

On the issue of scientific freedom, the young people divided into two groups themselves, the vast majority supporting the idea of social accountability. The few exceptions were—it was reported—the most scientifically ambitious types who argued, "To hell with it. I know what I'm doing, and I want to do it." There was a third group as well: those who didn't mind falling in with the public accountability gambit, because they were "pretty damn sure" that their own work wasn't going to fall within the area of interest. They were quite happy to vote with the majority.

One intriguing difference between the Americans and the Europeans was that, in general, the Europeans felt it was the proper function of governmental scientific coordinating bodies, such as Britain's Science Research Council, to determine questions of relevance, areas of responsibility and the biohazard rules that should operate in a laboratory. There was a marked resistance amongst the Americans, however, to face, let alone accept, the idea that the National Institutes of Health or any government agency should play that kind of role. In addition, all the Europeans, not just the British, accepted that the government through its agencies had the right to a determination of scientific priorities. The Americans were inclined to take the attitude that only institutional and peer-group committees could take such decisions, and initially they were very reluctant to let the National Institutes of Health or the National Science Foundation be officially involved *at all*.

At Asilomar, however, the officials from both those agencies had no illusions. They fully understood that they would have to be involved and that one couldn't discuss biohazards without acknowledging the rights and duties of the legal and legislative arms of the government. The problem was that scientists were trying to back both horses. If they admitted, as they did, that there was a serious hazard, then society and government became inevitably involved. But if there were no serious hazards, what was all the fuss about and what were they all doing at Asilomar anyway?

The cultural differences between the Americans and the Europeans on the point of political control probably find their explanation in the fact that, during the period 1945–65, American scientists saw the federal government primarily as a funding group with somewhat irrational policies, a group that funded research generously, but in a relaxed, noninterfering—one ungrateful scientist told me, "infantile" —way. In order to get funds, scientists had only to jump through a series of hoops. This state of affairs was delightfully exposed by Dr. Warren Weaver in a splendid, and now famous, tongue-in-cheek article, "The Report of the Special Committee on X," in which X stood for any piece of scientific research whatever. All that needed to be said—not proved— was that the work represented by X would help beat the Russians or would increase the military power of the United States or would have outstanding medical benefits or would bring a fantastic and lucrative technical spin-off. Then the grant came. The funding was far more liberal than anything their British or European colleagues could expect, but even when the scientists got it, they knew that there was an element of unreality involved. To say, as one scientist said, that "they had to lie through their teeth" is more than a little exaggerated. It would be kinder to say that they were forced into a "second-order truth," as Alan Bullock, the historian and Master of St. Catherine's College, Oxford, once put it.

In Britain, however, until quite recently, there was a genuine sense of a community of interest between, for example, the Medical Research Council, who gave the money, and the people who needed the money. By British standards, the funding was extremely liberal and rational: there was no unjustified interference; the scientists were closely involved in the discussions; the need for determining priorities was acceptable. Consequently the Medical Research Council was considered to be completely on the side of the angels. The people involved in the making of British science after World War II had very close ties with the MRC, and warm feelings towards it, as opposed to the United States, where few scientists were initially involved in the decision-making process anyway, and where those who were tended now to have bad impressions. Every American scientist has at least one horror story that he is happy to repeat about funding from government agencies.

The distrust of the agencies stems, I suspect, from the rapid change of scientific priorities initiated by pure politics. Over the last decade, a suspicion intensified that NIH had become an arm of the political branch, and the directorship of NIH a political appointment. In Britain, the Medical Research Council has rarely been closely involved politically, and to consider the director of MRC a political appointee would be regarded as ridiculous. The council is seen as the instrument of the scientific community which, if anything, speaks on behalf of the community to the government.

But whatever the scientists' past experiences or future predictions, the arms of the government did become immediately involved here. Less than twenty-four hours after the scientists had flown off, leaving the field clear to the migrating butterflies, a group of scientific administrators met in a hotel room in San Francisco, on behalf of the National Institutes of Health. They had the specific task of turning the vague, general conclusions of the Asilomar Conference into formal guidelines which, through their funding, would affect the scientists directly. The next chapter tells their story.

From the articles and the tapes, many of the flavors filter out; and they are in turn intriguing, amusing, disarming, inspiring. Berg's stature is one such flavor. Even though a few, a very few, people regarded the whole business as Berg's private obsession, his integrity was obvious and commanding. It is a measure of this stature, and a result of it, not only that, with one or two exceptions, the "pause" was widely observed, but that the conference could be held together *at all.* That a group of self-willed, self-interested, self-propelled, headstrong individuals unaccustomed to anyone telling them anything about their experiments was able to reach any agreement remains remarkable. But Berg, Brenner—indeed, all the members of the committee—knew that they had to reach an agreement on standards, or legislation would impose standards on them, as it may yet do. As Berg said, "If our recommendations look self-serving, we will run the risk of having standards imposed. We must start high and work down. We can't say that one hundred and fifty scientists spent four days at Asilomar and all agreed that there was a hazard—and then they still couldn't come up with a single suggestion; that's telling the government to do it for us." That injunction struck home.

The impact engendered by Sydney Brenner evokes another flavor. "Fantastic Sid," one scientist called him to me. According to Michael Rogers, he was "the single, most forceful presence at Asilomar," a "cross between a leprechaun and a gnome"—a cross made of recombinant DNA, no doubt! Everyone speaks with the highest praise of his contribution, though his tendency to railroad was sometimes resented, and like his colleague Francis Crick, he is not a modest man. He above all—except perhaps for Paul Berg—was the one person who had the clearest sense that the issue at stake was not one of scientific technology but one of political experience and political tactics. One scientist, recollecting the heated session on animal viruses, said to me: "I thought the whole meeting would collapse at that point because there was this tremendous antagonistic atmosphere. It was only people like Sydney Brenner who cooled the whole thing."

Then there were Watson and Lederberg, lofty and assured, as such giants usually are, arguing and squabbling with each other, for all the world like a pair of pernickety philosophers: Watson not really perturbed at all about imaginary dangers anyway, and leaving people uncertain about what he really felt about anything, except perhaps that the conference was serving no purpose, so had no point; Lederberg, very concerned about the possibilities of any kind of restrictive legislation, and right to be critical. Since there was no experience on which to judge anything, people were cheerfully making guidelines which didn't apply to *their own* experiments. But both of them, as one senior NIH administrator reported with delight, were being pushed aside by youngsters who had more exciting things to talk about, for the future was their's, not the old guard's.

Then there were those on the fringe: Carl Merril on the sidelines, excluded from the in-group with the unease of possibly unrepeated experiments hovering around his head like a cloud of germs; the Russians, isolated for other reasons: language and deliberate self-effacement and their own worries about government intervention in science. With the memories of the Lysenko era as a nadir of Russian science, the Russians reportedly were initially furious with the Americans for bringing the issue up at all and so opening the door for government interference again.

There were the flavors of moment: humor, when one scientist realized that according to their classifications, sex was a highly dangerous experiment that should be banned; nasty disagreements of such intensity that another young British scientist would say: "The whole thing was a disaster. Nothing really was achieved. Nothing except to show the astonishing spectrum of opinion from individual scientists. We were not struggling at all to get to grips with the problems of the rest of society and I didn't like our attitudes. Who is to know that an experiment with cold-blooded vertebrates is more safe than with warm-blooded, on a show of hands amongst a hundred and fifty scientists? This is just nonsense." Another scientist said to me: "One of the most distressing things about the meeting was the lack of acknowledgment of this factor [public accountability]. For what the meeting was looking for was public approval for scientifically determined priorities, rather than accountability to publicly determined priorities. They would make the experiments as safe as they possibly could—having done that then everything could go on just as before."

But in fact, all this is too negative: something tremendously important was achieved. The process of self-education that began there has continued. The message finally did come through: the public may have, and may choose to exercise, the ultimate lever on what research it regards as worthwhile—and it may not have the same priorities as the scientists. This factor would never again be forgotten.

7
Creating
New Guidelines

*Ye blind guides, which strain at a gnat, and swallow
a camel.*

—Matthew 23:24

Just over twelve months elapsed between the Asilomar Conference and the publication of the final NIH guidelines to which laboratories must now conform if they are to conduct recombinant experiments under grants from the U.S. federal government. With slight variations, these guidelines also form the basis for the regulation of experiments in Europe, especially those experiments conducted under the aegis of the European Biological Organisation.

Those months were not entirely happy ones for the scientific community at large, nor for certain people within it. In some instances, slight differences of opinion which first appeared at Asilomar widened into rifts, if not chasms, as the scientific and social issues became more tendentious. The arguments still continue over such questions as how dangerous the procedures are, and therefore just how careful scientists need to be; exactly how sanctions can be applied against those who ignore the regulations; what degree of policing of scientific laboratories is either called for, necessary or tolerable; just what degree of public participation is appropriate; and precisely what the relationship of scientists with the politicians of Congress and the administrators of the central government should be. It would be an exaggeration to say with the Bible that "nation was set against nation, and brother against brother," but such has been the depth of the issues that some scientists who were natural allies over such disparate issues as Vietnam or scientific methods find themselves again in an atmosphere of charge and countercharge, questioning one another's motives and impugning one another's beliefs.

Once again, the form of the discussions and their outcome vary between America and Britain, and even between America and the rest of Europe. This is due to culture, of course, but also to the facts of history. For Britain, the drama is being played against a backdrop which contains two strongly disparate elements, both of which are missing from the American scene: a recent accident with hazardous viruses which resulted in four deaths, and a very powerful trade-union movement, concerned with the fullest protection for workers against everything. By contrast, the American tapestry contains the intense competition of scientific research, with science itself almost an industry by now, together with a remarkable degree of willingness to examine questions in the open, and under public scrutiny. These open, wide-ranging discussions are inimical to many Europeans— scientists and civil servants alike—who play down what one European scientist unfairly described as "the usual kind of American fuss."

Of course much of the heat would be taken out of the disputes if we knew or could establish the danger of the experiments. This is very important for the general safety of future procedures and also because, as was shown in Chapter 2, the experiments themselves are scientifically so valuable and fairly easy to do. The enzymes can be bought for about four dollars a bottle from the Miles Laboratory at Elkhart, Indiana. Cultures of *E. coli* can be bought from Types Culture Collections, and with this fairly modest outlay, you are in business. While it is far simpler and cheaper than making an atomic bomb, it is still unlikely that any problems will arise because of the deliberate action of some adult or even adolescent Frankenstein. If problems occur at all, they will more likely result from people's carelessness or thoughtlessness or indifference to accepted safety procedures.

It should never have happened at the London School of Tropical Medicine and Hygiene, but it did. Of all the places in Britain, this laboratory has great experience in dealing with very dangerous organisms, and you would expect them to have ironclad procedures for isolating any of their workers who seemed to be developing anything suspicious. But in 1972 one person was careless and contracted smallpox. Unfortunately the symptoms were not unequivocally obvious right off, and she was placed in a public ward in a general hospital. The parents of another patient, coming to visit their daughter one afternoon, sat on chairs between the two beds. All three—the original contact and the visitors—as well as a nurse died of smallpox. How could it have happened? Old equipment, old buildings, procedures which presumably were never updated, carelessness about security and human error—all five elements contributed to the tragedy. "They had it coming to them" was the unsympathetic remark that someone made to me at the Imperial Cancer Research Fund, concerned about their own safety regulations.

Accidents of this kind do happen at times. Here was a known hazard, a dangerous pathogen, and yet not only were inadequate vaccination methods used, but people were allowed to wander in and out of the laboratories quite casually. No one identified the case of smallpox when it was contracted. It was only by the sheerest accident that it was picked up at all. Even then, the safety committe at the college lacked awareness of what was happening. How was this possible? The episode could be portrayed as demonstrating a total lack of scientific or social responsibility, but in reality it simply resulted from pure carelessness and boredom, and that is how these accidents really happen. Even after all the NIH recommendations and fuss I found it ludicrously easy to wander in and out of the laboratories in Britain, Switzerland and America at will. I remember coming into one laboratory through the back door, by accident, since I couldn't find the front, and being waved through the forklift entrance. I was then able to wander around the corridors posted with signs saying NO ADMITTANCE and VISITORS PROHIBITED, everyone waving me on toward my destination in the friendliest of fashions.

During the last twenty-five years there have been four hundred and twenty-three infections and three deaths from dangerous viruses at Fort Detrick, Maryland, a maximum-safety center. Someone once picked up an aerosol container and found he had sprayed the laboratory with plague. Recently a worker at Porton Down, also a maximum-safety center, contracted Marlburg disease from a highly dangerous virus when his protective glove split. There have also been serious accidents at the National Institutes of Health; there are not many of them and they are not publicized, but they almost *always* result from breaches in laboratory technique. To take an example, there are standard rules in England and America for the use of radioactive isotopes in laboratories. In America, the Atomic Energy Commission has policing authority, and all new isotopes have to be bought from them under license. If the rules are not followed, the purchasing license is canceled, but it is not always possible to tell if the rules are being followed. Radioactive fluids are an indispensable tool in contemporary biological research, and it is forbidden, for instance, to suck them up by mouth. Consequently, some rather complicated hand pipettes have been developed, but they are a nuisance to use and don't work so well. If no one is looking, a scientist or technician is often tempted to measure the material by mouth pipette. One scientist told me that when he took charge of his new laboratory, he suspected that something like this might be going on. He asked everyone for a urine specimen to make sure they were not receiving too much radiation exposure. They all provided it willingly except for one person who became furiously angry, accusing the scientist of unjust interference and an intolerable invasion of his privacy. My friend insisted, and finally the man went off, took his own sample and tested it himself. He stormed back into the laboratory, in a towering rage. "I thought he was going to hit me," said the scientist. "He said, 'That's a really dirty trick, giving a guy a contaminated test tube.' He had found a large degree of radioactivity in his urine, and he

thought I'd pulled a fast one on him and given him a contaminated specimen tube. He was livid. One of the other
people in the laboratory intervened at this point and said,
'Come on, go take your own perfectly sterile tube and do it
again.' He did just that and returned white as a sheet, asking
if anyone knew a good attorney. I asked him why. Did he
want to sue me for invasion of privacy? But it wasn't funny
at all. He had a very high level of radioactivity in his body
and said he'd better get round to making his will. He must
have been mouth pipetting for months."

Radioactive hazards are proven, however, and those of
recombinant DNA are not. DNA itself has been around for a
long time, and human beings have been exposed to it for
eons. Can foreign DNA integrate itself in human cells and
cause the kind of havoc that ultimately gives rise to cancer?
In what form, if any, is DNA dangerous, and are the recombinant experiments as such likely to present it to human beings in a dangerous form? Probably not.

First of all, as Dr. Sydney Brenner reminded me, infectivity normally cannot be detected in most normal kinds of
bacteria but only in very special varieties and under very
special conditions. Moreover, naked DNA, without its coat,
loses infectivity. If one could do experiments solely on that
basis—with only raw DNA coming in one end and raw DNA
going out the other—the chances of infectivity would be
reduced by a factor of millions. "We won't, of course," Dr.
Brenner insisted, "slosh the stuff around; it will be handled
carefully; it will be decontaminated." But it might not be
possible to design all experiments so as to avoid handling
material which in nature has been designed with a special
infective apparatus.

Another point is that phage by itself is not necessarily
dangerous. The experiments at the Pasteur Institute in Paris
during the 1940's and 1950's that laid the groundwork for
knowledge of the gene used phages picked out of the Paris
sewers for experimental material. At Harvard Medical
School, the students of Dr. Richard Goldstein examined the
phages in the local sewers as an experiment. It was a kind of
scientific treasure hunt: go into the sewer and see how many
phages you can retrieve. Every time we chew a hunk of

anything, we take in—or more accurately, expose our intestines to—bacterial and phage DNA in the raw, and we have
been doing that for millions of years. And even if cooking
doesn't kill off much of the DNA, there are known to be
powerful intestinal enzymes that destroy it.

However, it is not enough to assume that the bacteria with
the recombinant DNA will not flourish in the intestine; a
scientist must show that it does not penetrate the gut wall
and pass from the intestine into the body. A number of
scientists, in the time-honored tradition of testing things out
on themselves first, have been swallowing some of these
plasmids and bacteria, and examining their feces afterwards
to see if the plasmid is still around. Dr. Kenneth Murray
revealed at Asilomar that he had swallowed his own favorite
phage, *Lambda,* and had been unable to detect it subsequently. Dr. Charles Weissmann of Zurich went one step
further in his attempt to see whether a recombined plasmid
stayed around in the gut. He began with a strain of *E. coli*
that was streptomycin-dependent. Since it needed an environment of streptomycin in order to prosper, it couldn't do
any harm if it got loose in his body. Into this *E. coli* he
inserted a recombined plasmid, with a gene resistant to another antibiotic, kanamycin. He cultured up the bacteria till
he had about one gram in weight of bacteria—two hundred
thousand million organisms, give or take a few million!—put
the culture in a large glass and swallowed the whole,
unappetizing mess. It tasted awful.

Then he tested his feces for the plasmid every day; after
three days he could not detect it at all. This might have been
a clear-cut answer, but for one other thing: though he could
not detect the plasmid containing the gene resistant to kanamycin, he did nevertheless have a population of *E. coli* in his
intestine, which *was* resistant to kanamycin and which he
apparently didn't have before the experiment began, or if he
did, it was so small as to be undetectable. How did it get
there? Had the gene for kanamycin resistance been somehow transferred at a local level, from the plasmid to the *E.
coli* population already in his intestine? The experiment and
testing went on for three months, and though there was
never *any* trace of the plasmid, the kanamycin-resistant

strain of *E. coli* persisted. But one day at the height of the cherry season, Dr. Weissmann, who I infer must be very partial to fresh Swiss cherries, scoffed a kilo or so of them and inadvertently gave himself a nice little case of local diarrhea. The unwanted population of *E. coli* vanished within forty-eight hours, and the normal flora and fauna of Dr. Weissmann's intestine were restored.

He repeated the experiment, with the same results. His conclusion: the population originally swallowed did not stay around, nor was the plasmid transferred in the intestine from the swallowed *E. coli* to another group naturally residing there. Instead, his swallowing of that vast quantity of bacteria had temporarily changed the whole ecology of his gut—until the cherries changed it again—and this allowed a small group of kanamycin-resistant *E. coli, which must have been there anyway,* to predominate and flourish for a short time. This was not surpising, for as Dr. Weissmann insisted, it is very difficult for a new population of *E. coli* to establish itself when the ecological niches are already occupied—as they usually are in a healthy gut. A new population can be rinsed out comparatively easily, for over time, established populations evolve complicated cell surfaces which enable them to cling to the walls of the intestine in which they are living. All available space is occupied, and it therefore requires a new, better-adapted population to dislodge them—better adapted, that is, to the gut wall to which they are attached; or another factor, like an antibiotic, to loosen their grip and flush them out so as to leave the space available for colonization by another group. Though this experiment by itself is not conclusive, it does appear that whatever the change induced in Charles Weissmann's intestine, recombined bacteria did *not* incorporate with existing ones, and they did disappear fairly easily.

The significance of such experiments is debatable, however, and arouses the usual kinds of disagreements amongst scientists. One rather disgruntled NIH scientist, commenting on Ken Murray's act, described it sourly as "a typical piece of British arrogance." He insisted that swallowing the recombined phage proved nothing, for it was still possible that the DNA of the recombinant molecule had invaded the intestine. Perhaps Ken Murray has *Lambda* somewhere inside his cells, and Charles Weissmann has his recombinant DNA inside him, too.

There is another test that could be devised to check this, but it brings up a question which has been hotly debated ever since Asilomar. Should one do the "Dangerous Experiment" and try to settle the question once and for all? That is, should one take some DNA *known* to be dangerous, construct a hybrid plasmid, stick that into *E. coli,* feed it to an experimental animal, and see if the plasmid—or any part of it—turns up anywhere in the animal's body?

If the infection should get loose, it could wreak havoc. It would have to be done in a situation where there were exquisitely elaborate physical containment facilities. At the present time this experiment is being designed at Porton Down in England, with the collaboration of the Murrays of Edinburgh and Charles Weissmann, under the auspices of the Europen Molecular Biological Organisation, and a similar one is being planned by Drs. Martin and Rowe at NIH. The experiment described below is the European one, and not only are the Murrays of Edinburgh involved, along with Charles Weissmann; so also is the Imperial Cancer Research Fund. It takes several steps.

Step one will be to take a newborn germ-free mouse and to infect it by injecting under the skin and dropping through the nose, ears and throat the DNA of the polyoma virus, a highly dangerous tumor virus. The mouse will become infected with this tumor virus, and the virus will turn up and replicate in the kidneys, amongst other organs.

Step two will be to make two recombinant DNA hybrids for subsequent infection: hybrids of a plasmid and the polyoma virus, and of *Lambda* phage and the virus. Then the experiment will be repeated, but this time the mouse will receive the recombinant viruses, in the same manner as before. Again, scientists will expect to find traces of the infective virus throughout the animal's body.

Step three will be to infect species of *E. coli* with these plasmids, but now *feed* them to the mouse, instead of injecting them, and see whether this *E. coli* lodges in the intestine of the mouse at all, and if it does, see what happens to it. Will hybrid DNA be exchanged and penetrate, and will it end up in the kidney tissues? If the results of the experiment are negative, which Ken Murray thinks they probably will be the first time round, it will have to be done again in more refined detail until some kind of positive answer seems to emerge.

The scientists are now busy devising the protocols. Newborn mice will be infected with polyoma virus at the Imperial Cancer Research Fund in London. The construction of the recombinant plasmids will be done at Edinburgh and Zurich. But the actual experiment of putting the hybrids into *E. coli* and infecting mice again will be done under the strictest conditions of physical and biologial containment (known as Category P4) at Porton Down in England.

If after all the effort, which will probably be long and drawn out, the mice do come down with infective polyoma, and traces of the recombinant DNA *are* found in the organs of the mouse, then at last scientists will know where they are: DNA *can* be assimilated through the intestines, and the experiments *are* dangerous. Nothing is quite safe, not even Ken Murray's gut nor Charles Weissmann's. The rationale of the experiment will have more than justified itself: to persuade people to follow the safety regulations. If scientists do not really believe that the experiment is dangerous, they will be careless. Most scientists working in this area continue to feel that the risks are no greater than those normally attendant in established bacteriological and microbiological laboratories, and only a positive experiment, with positive results, will convince them. This, not policing the laboratories, will make them stick to the rules.

If there is a negative finding, though, there will be a problem, and this is where some scientists—Maxine Singer is one of them—are worried about this "Dangerous Experiment." They are concerned that a negative finding will be interpreted as meaning that all such experiments are now totally safe, that people will now think they have a green light— whereas of course all that a negative finding from *this* particular experiment will tell is that polyoma virus DNA from a recombinantly infected *E. coli* did not get into a mouse. There might be any number of reasons for this which, whether the outcome of the experiment is positive or negative, would not apply to man. It will be repeated with different strains of *E. coli* and different viruses, but while one single experiment may show the danger, no one series of experiments will unequivocally show the safety—now or ever. What Dr. Weissmann thinks will happen is that over a period of time, the sheer weight of negative findings in such experiments will suggest that plasmids are *not* transferred through the gut wall from bacteria, and therefore that the routine is not highly dangerous. But for the moment scientists are caught between the Scylla of continuing ignorance and the Charybdis of continuing uncertainty. And of course they cannot perform the crucial, the "*Absolute* Dangerous Experiment." As with cancer viruses, they cannot do the experiment on themselves or any other human being, for who is going to take the risk of being infected with a highly dangerous agent like a polyoma virus?

This dramatic experiment is one method being designed to resolve one outstanding issue. Since Asilomar. however, other forces have been at work, as well, with the aim of controlling recombinant DNA research—the regulatory commissions of the British government and the NIH. In England there has been a quite deliberate intention to play the whole situation "cool"; a desire which has filtered down through the corridors of power ever since the first worries were raised, occurring as they did so soon after the smallpox outbreak at the London School of Tropical Medicine, which was the subject of a public inquiry. By the time the moratorium was first announced, the British Medical Research Council had already sent a confidential letter to all directors of laboratories who were receiving grants, banning all experiments that were affected by the voluntary moratorium elsewhere. The Scientific Research Council did the same, and it is a piquant element of the situation that neither Dr. Kenneth Murray nor Dr. Noreen Murray ever received such a letter. Ken Murray is still highly amused by it all. There was, he admits, no reason why MRC should write to him, since he was not receiving grants from *them,* but his last grant proposal *was* read in the corridors of power of the Scientific Research Council. Someone, however, came to the conclusion that his experiments were not germane to the issues of recombinant DNA! They were—very—and the Murrays observed the moratorium voluntarily.

After that, two commissions were quickly set up to investigate the dangers of laboratory research, both because of the DNA furor and because of the smallpox outbreak. The first commission, under Lord Ashby, came in for a lot of sniping; the working party was made up entirely of scientists, and in striking contrast with the NIH inquiries in America, it did not involve the public in any way—nor has it really yet. Moreover, there is a feeling that these "in-house" committees in England tend to become very chummy and relaxed, with little sense of urgency or of priority and a quite discernible touch of "laissez-faire." Some suspicions were voiced that it was just a public relations exercise, and that the liberal interpretation of the phrase "work involving potentially serious hazards," the subject of the second commission under Lord Godber, was quite subjective, indeed quite accommodating. But the final recommendations turned out to be quite the opposite.

But early in 1975 a third working party, under the chairmanship of Sir Robert Williams, director of the Public Health Service Laboratory, London, was instructed to come up with a strong code of practice for such research and to make recommendations for the establishment of a central advisory service for those laboratories which would use the techniques of genetic manipulation. In announcing this, the U.K. Secretary for Education and Science, Mr. Kenneth Mulley, said that they would be in full consultation with all concerned, including management and trade-union interests.

At the same time, on the other side of the Atlantic, the Recombinant Molecules Advisory Committe of the National Institutes of Health had gone into action, meeting in San Francisco as soon as the Asilomar Conference ended. No one from the Asilomar organizing committee belonged to the NIH committee full time, though Maxine Singer, Paul Berg and Sydney Brenner were consultants. Berg told me later he decided not to sit on the committee, because he felt it was important that people with vested interests in the outcome should now start to become detached. This was a wise move. The problem before the committee was this: Insofar as safety regulations were concerned, how should NIH handle the expected flood of grants for recombinant DNA research? The committee first settled some simple procedural matters, such as that every applicant institution should have an advisory committee of its own to review the physical aspects of the laboratory and the educational background of the workers, and that if the institution were too small, a neighboring one should supply the service.

They then sat down to consider the experiments themselves: what they were, how they should be classified by hazard, what kind of containment was required. Initially they considered physical containment only—hoods, boxes, rubber gloves, negative air pressure and showers—but they quickly became intrigued with, and impressed by, the possibilities of biological containment: using organisms so designed that their survival would be conditional upon unusual thermal or chemical enviroments. These would provide a great deal of assurance, for they could not possibly cause an epidemic. Though at Asilomar the idea was just a gleam in the eye, the possibilities of biological containment have expanded greatly since then.

They set up a subcommitte chaired by Dr. David Hogness of Stanford University to write draft recommendations, and their final report combined both methods of containment in a complex formula that depended on the degree of danger of the experiment. In a final meeting in Woods Hole, Massachusetts, in July 1975, these proposals were discussed by the committee, with plenty of harmony and agreement, possibly due to the fact that many of them were not present! The chief bone of contention seemed to be whether people sitting around the table should be allowed to smoke. They agreed fairly easily on some guidelines which were edited, written and rewritten, redrafted and redistributed. This harmonious meeting did something else as well, however; it weakened the proposed guidelines relying more on physical containment so that the final ones were weaker than those suggested by Hogness's subcommittee and than those suggested by Asilomar.

They expected no difficulty—but then the angry letters started to come in. The first criticisms came from Paul Berg. He was agreeable to the Hogness draft but not to the watered-down, Woods Hole version. He felt all methods of physical containment were quite overrated so far as safety was concerned. He thought that much greater emphasis should be given to biological containment, a view widely shared by his colleagues in Europe. With the evidence of past accidents before them, they felt that the only really safe way to proceed would be to concentrate on "self-destruct" bugs.

More critical grapeshot came in the form of a petition signed by fifty people, organized by Dr. Richard Goldstein of Harvard Medical School and Dr. Harrison Eccles of the University of California at Berkeley, who raised several objections. They pointed out that the guidelines *were* lower than those accepted by the community at Asilomar; they felt that *all* shotgun experiments using mammalian tissue should take place under the strictest conditions of physical containment (P4); they felt that the constitution of the NIH committee was totally inadequate, and that there should have been more representation of the public at large, a wider range of representatives from other scientific disciplines, and most importantly, more members who had no direct interest in the outcome. Lastly, Richard Goldstein raised fresh doubts about the safety of *E. coli,* feeling that it should not be used under any circumstances. Priority, he felt, should be given to devising a new bacterial host which *cannot* infect man. Though the laboratory strain of *E. coli* is considered feeble, even those who work with it admit that not enough is known about the ecology of this bacterium and its associated stages. Since it infects man with a childish ease, Goldstein considered that the choice of that bacterium was "reckless," even if it was a natural one, scientifically speaking. Finally, the petitioners recommended that there should be a postponement of the most hazardous experiments: things could be held in abeyance for at least five years and it wouldn't make all that much difference. Certainly the scientific world would not come to an end!

Another broadside, through press releases and conferences, came from the Genetics and Society Group of the Harvard Branch of Science for the People. Through their spokesman, Dr. Jonathan King of the Massachusetts Institute of Technology, they also pressed hard on the vested interests involved. King argued that the committee, as it was formed, protected geneticists, not the public, and having, as chairman of the subcommittee that wrote the guidelines, an active worker in the recombinant DNA field, was like having the chairman of General Motors write the specifications for safety belts in cars.

If the storm of reaction was heavy, though, the storm of counterreaction was enormous. The NIH committee took the greatest exception to the comments attributed to the Goldstein petition about vested interests. Scientists were absolutely furious that the issue had been brought up at all, especially in the form of a petition. Goldstein was told privately that he "ought to keep his mouth shut—not rock the boat," and there were murmurings about "left-wing elements who were trying to destroy science." The Goldstein group was very upset about this, especially since many of the Cambridge scientists had taken an active stance over the Vietnam War and were known for their somewhat radical attitudes. They resented what they felt was a nasty smear tactic, for the suggestion that they were "out to destroy science" was sheer nonsense. Yet others accused the Harvard group of rank hypocrisy, saying, "You should see the experiments some of them are doing!"

Why were so many scientists so suddenly on the defensive? It is difficult for an outsider to be absolutely certain, since as reporters of science have often stated, in an explosive atmosphere of charge and countercharge, defense and aggression, scientists will say many things as background, but you may *not* quote them by name. But it really does seem that there was a genuine division on an old point: public involvement. A number of people on the committee were defensive because they felt that unless there was decent public scrutiny of what was proposed, the safety standards would not be adequately applied. On the other hand, there were one or two "cowboys"—scientists who wanted the laxest controls possible and who made their presence strongly felt at the Woods Hole meeting and a later one at La Jolla. Others, such as Drs. Stanley Falknow of Stanford and Roy Curtiss of Alabama, were quite concerned lest the whole committee be regarded as a façade. Experiments could easily go on in situations which were not at all safe, especially in the industrial arena, where vast financial potential was involved, and where the long arm of the NIH regulations did not extend at all. (It still does not.)

People such as Dr. Roy Curtiss, whose team in Alabama had spent a most exhausting year doing their utmost to design a safe vector, were also shocked at the implication of the "cowboys" that this was already to hand. It wasn't at all, but so anxious were scientists to have such a vector that someone almost had only to stand up and announce the fact for everyone to agree!

As a result of all the pressures, two things happened. The Goldstein group was *not* invited to give evidence at the next meeting of the recombinant DNA committee, which met at La Jolla in California in December 1975, and the chairman of the committee, Dr. DeWitt Stetten of NIH, decided that maybe they ought to take another look at the guidelines. He told me that as the pendulum swung between the scenarios painted by the scientists, he found himself alternating between terror and laughter. Another small group was formed, chaired now by Professor Elizabeth Kutter, to collect opinions from some of the dissidents and frame a third draft of the guidelines. By the time the committee met at La Jolla, they had three drafts to consider, and they were forced to steer through very choppy waters, to maneuver between the laxity demanded by impatient biologists on the one hand, and on the other the danger of imposing such restrictive rules that people would flout them and privately slip into the laboratory during the weekend to do what is known as the "Saturday-night experiment."

Before any final decision was made, however, and in response to mounting criticisms and an increasing public interest, Dr. Donald S. Fredrickson, director of NIH, decided to call a public hearing in February 1976, almost one year after Asilomar. The ultimate responsibility for deciding just how strict the rules should be would be his. The hearing was conducted by a twenty-member advisory group, including the chief judge of the District of Columbia Court of Appeals, David L. Bazelon; the former general counsel of the Food and Drug Administration, Peter B. Hutt; and the current president of the National Academy of Sciences, Phillip Handler.

The committee found no difficulty in grasping the important issues, despite the scientists' arguments that the levels of technicality were such that they could really only be judged by experts. The details of the procedures whereby the sticky ends of DNA are made to recombine are probably no more intricate than those of a complicated murder case. Again, the pressures came on the question: are the guidelines too lax or already too strict—or just about right? The group from Harvard Medical School, who felt that they were too lax, was represented by Richard Goldstein and Alan Silverstone from the Massachusetts Institute of Technology. Having no vested interest in the outcome, they were, amongst the scientists, in a psychologically and morally strong position as they reemphasized their earlier critique.

On the other side, scientists such as Donald Brown of the Carnegie Institute of Washington, David Hogness and David Baltimore all argued that the guidelines were possibly already too strict. Baltimore said that, so far as tumor viruses were concerned, "It is only barely possible to go forward." The usual arguments raged back and forth.

The committee took written evidence also. The most interesting, the most reflective, and what may turn out to be the most crucial came from Dr. Sinsheimer, chairman of the Biology Division at the California Institute for Technology, who, as has been seen in Chapter 4, once held strong views about the worth and possibility of genetic engineering. He now believes that no research of this nature should go on, unless it is confined in safe places such as Fort Detrick in America and Porton Down in England. His arguments were by no means negligible. They were new in two respects. First, he was prepared to take a look at the evolutionary and biological question in all its complexity, rather then in the narrow context of one area of research technology. We must remember that it was a failure to look at—or appreciate—the wider complexities that resulted in a program of hormone therapy in pregnancy, which in the early 1970's was found to have the appalling consequence of causing vaginal cancers in the young daughters of those who took the hormones. Most simple scientific steps *are* perfectly logical and seem quite justified. In this instance, it was noted that in pregnancy estrogen levels rise. So people who have had consistent miscarriages could, it was argued, be treated by maintaining the levels artificially. Estrogen is a natural substance, too, and therefore could do no harm. What scientists did not—*could* not then—appreciate was that hormones work in such delicate balances that while the right hormone at the right time can do wonders, the right hormone at the wrong time can be disastrous. Similarly, Sinsheimer wanted to say that there is simply so much we don't know about the total effects of recombinant DNA that something might indeed go wrong.

The second element in his argument was this: suppose there is and so far this is neither proved nor refuted—a barrier to genetic exchange between the two classes of living things. Suppose it is true that the primitive cells of bacteria and blue-green algae, which do not have a nuclear membrane, cannot exchange genetic material with the infinitely more complicated cells of all other higher organisms which do. (One piece of evidence for the existence of a barrier against such genetic "intercourse" is the completely different signals the two classes maintain for determining the timing operation of their genes: the actual genetic code may be universal but the elements that control it are not.) So, Dr. Sinsheimer argued, the danger of putting a piece of DNA from a higher organism into a lower one lies in the fact that thereby one may be handing over control signals along with the DNA—what Nicholas Wade called, in a most felicitous phrase in an article in *Science,* "a sort of betrayal of state secrets at the molecular level."

Well, would this matter? Sinsheimer thinks it would. One result of that "betrayal" might be that certain viruses which cause bursting damage to bacteria but which normally do not infect higher cells might now acquire that capacity, and cause the cells of higher organisms to "lyse." Another possibility is that bacteria might become pools—positive reservoirs, in fact—of common viral infections. Sinsheimer is making two points here: not only are biologists on the verge of creating new life, but *by this very act* they will disturb an extremely intricate ecological interaction which at present is only dimly understood. In the same way that endocrinologists disturbed an extremely intricate hormonal interaction in women's bodies, not through Machiavellian ploys, but through well-meaning ignorance, so biologists could be disturbing a most delicate balance, and through the same kinds of motives.

He is not arguing for stopping the experiments, only for confining them. But to have to go to Fort Detrick or Porton Down would be another intolerable restriction for some scientists. As Sinsheimer pointed out, neither he "nor anyone else can say that if the present committee guidelines are adopted, disaster will ensue. I will say, though, that in my judgment, if the guidelines are adopted and nothing untoward happens, we will owe this success far more to good fortune than to human wisdom." Though there is no evidence yet for his belief and though some scientists just don't believe him, he possibly could be right.

On June 23, 1976, Dr. Donald Fredrickson finally issued the guidelines. They ended up stricter than both those recommended by the Asilomar Conference and those recommended by the committee that advised him. Dr. DeWitt Stetten described them as "in no way opening the floodgates —rather [they are] a closing of leaks [in the Asilomar guidelines]."

The guidelines called for two types of containment. We can best think of them as two lines of defense. The first line consists of physical containment, ranging from the loosest conditions of experiment (P1) to the most restricted and secure (P4). P1 corresponds to the standard pattern of microbiological techniques: experiments are conducted on open bench tops and special equipment is not required, but work surfaces are regularly decontaminated, and eating and smoking, while not forbidden in these laboratories, is not encouraged. The restrictions get increasingly tough until, by the time P4 is reached, the laboratory is a completely isolated and controlled facility: one enters and exits only through airlock systems; discarded biological material and the special clothes which must be worn are sterilized before removal; all outgoing air is filtered; the whole laboratory is kept at a negative air pressure compared to the outside air, so the air inside cannot escape; access is strictly limited to those taking part in the experiment; and all procedures must be done in biological safety cabinets.

The second line of defense is the biological one: using only suitably weakened strains of *E. coli.* An EK1 experiment permits the use of the standard strain that has been used for decades for genetic experiments, but an EK2 strain must be one which, according to laboratory tests, has a less than one thousandth of a millionth chance of surviving outside the laboratory, making the chances of its establishing itself outside remote. The EK3 category is the same, except that the survivability of the bacterium has actually been tested in plants and animals and external laboratories as well.

Do such highly weakened strains exist? The EK3 strain does not yet, but the EK2 strain does. Dr. Roy Curtiss of the University of Alabama and his team have produced one, labeled in honor of the Bicentennial, number: 1776. Two plasmids are built into this strain, one of which is our old friend, Plasmid Stanley Cohen 101. This combination virtually cripples the bacterium for survival outside the laboratory. These categories then are expected to take care of all recombinant DNA experiments, particularly the most hazardous ones, which will be governed by a combination of P4 and an EK3 strain of *E. coli.*

When I read the guidelines I could see both the reason why Peter Carlson could take me only part of the way with our experiment with human insulin, and why scientists are not unduly anxious about the imminent arrival of wholesale human genetic engineering on the scientific scene. For the guidelines really do restrict the kinds of experiments that can be done in this area at present. For a start, all shotgun experiments involving human DNA must be done either in P4 facilities using an EK2 strain, or in P3 containment using an EK3 strain. Since this latter strain does not yet exist, this means, in practice, that these types of human experiments can only be done in those few places, like Fort Detrick, that have the most rigid types of containment.

Some experiments are still forbidden outright: those that seek to clone DNA from known pathogenic organisms or known cancer-causing agents or genes that code for potent toxins, such as snake venom or diphtheria; those that involve the deliberate creation of pathogens that would be likely to increase the virulence of *anything;* those that would result in recombined DNA being released into the environment;

and those that would bestow drug resistance on a microorganism that normally would not acquire it naturally.

These guidelines are tough and were meant to be. Dr. Berg told me that he was happy with them. Though he thinks they are tougher than they need be, this is, he insists, an error on the right side, since he still feels that it is essential to lean over backwards, not only to be safer than is perhaps necessary, but to be seen to be safer. It would, however, be too drastic—in his words, "quite unacceptable"—to wait until there was a *totally* safe vector. As it is, the NIH guidelines are only *just* acceptable to the Europeans. Dr. John Tooze, the director of EMBO, has gone on record as saying that, had they been one jot tighter, the Europeans could not have accepted them. Admittedly there is no onus on the Europeans to accept them even now, but they have, for among other reasons, everyone would be unhappy with a situation in which there was such a drastic difference in procedures between countries that the whole thing became a farce. As it is, I heard from three different but equally impeccable sources in Britain and Switzerland that people had already gone to Switzerland to do certain experiments which in the interim they could not do at home. Ray Dixon of Sussex told me that every time he visits Hungary, he is teased mercilessly, since his friends say the purpose of his visit is to do experiments he can't do at home! This happens not to be true, but there is no doubt that scientists would up and go if they felt the restrictions were intolerable.

Some four months after the NIH guidelines were issued in the United States, the Department of Health and Social Security published proposals for statutory control over genetic manipulations in Britain. These were based on the Williams report. In many respects the suggested controls in Britain are far more extensive than any proposed by the NIH, since they will cover not only university science laboratories but those in industry too, whether the industries are concerned with brewing or fermentation or pharmaceutical processes. In addition, the controls encompass a wide variety of both *teaching* and research facilities in microbial genetics, in medical and veterinary institutions as well as in traditional university laboratories. The response from some quarters has been fierce, using amongst other formats that most loved one: "the

letter" to the *London Times*! For example, Professor Pirt of
Queen Elizabeth College, London, called the recommendations "the first outright attack on scientific freedom in Britain." He feared that the bureaucratic superstructure that
would be necessary to enforce the very wide blanket regulations could have the effect of "trivializing" microbial research in Britain, just at a time when the subject needed
every encouragement, because of the contributions it could
be expected to make to, among other things, the nation's
health. Other groups, too, were angered by the breadth of the
regulations. They believed them to be thoroughly unwork-
able, since they would require, among other things, advance
notification of many experiments that are currently not considered hazardous, including some demonstrations that
actually take place in secondary schools! Between depart-
mental proposals and parliamentary legislation, however,
much time and agony will be involved, so as this book goes
to press, it is impossible to say what the final form of the
regulations will be.*

In Europe, the debate seems to be much quieter, even non-
existent. The work in such places as Charles Weissmann's
lab in Zurich is controlled but not so restricted, and there is
nothing like the same degree of public concern.

Wide public debate has been restricted to America. European scientists are clearly deeply thankful about this and,
holding their breath, are anxious that such debate should
never surface. When I suggested to one British scientist who
now works in Germany that the British public should have
been more involved, or at least have been given the option
to become more involved, I received the avuncular and
deeply self interested reply that there were many other more
important matters of deep principle they should be discussing, such as constitutional reform. But clearly he was not
about to initiate any public debate about science or politics.
The situation in France developed in its own charmingly
French manner. It will be recalled that there was a furious
debate at the Pasteur Institute in July 1975 about whether
the work should be permitted there at all. July is followed

---

*A genetic manipulation advisory board, made up of nonscientists, now
keeps a watching brief on the research.

by August, and in August Paris empties as everyone, of whatever opinion or political hue, disappears for the summer holidays. In that month the late Jacques Monod quietly arranged for containment facilities to be built in his own personal laboratories of molecular biology, across the road from the main institute. The dual impact of a summer holiday and a *fait accompli* effectively stilled the arguments, especially as "across the street" was not deemed to be so dangerous as "under our roof." The work goes on there to this day.

In Russia a debate has been in progress since 1970, according to Victor Zorza, the famous U.S.S.R. watcher. It has taken on its own special flavors too. There have been similar differences of opinion about the dangers of recombinant DNA research, and disagreements as to whether it should be restricted. The debate has also extended to the problems of cloning and genetic engineering in man. That there has been any debate at all should give some comfort to those who fear that a totalitarian government might exploit these possibilities wholesale. Many Soviet scientists have backed away from genetic engineering and cloning in man, on moral and social grounds, but others of the top Soviet scientific elite have urged that the social improvement of man should be speeded up by all biological means possible. Many of the arguments parallel those in the West, and so do many of the questions: Is science neutral? Who should control it? In June 1974, however, the party issued a decree calling for a wholesale expansion of molecular biology. Zorza interprets this as a victory for the director of the Soviet Institute of Biology, Vladimir Engelgardt, who had struggled long to get the party's backing. He is quoted as saying: "[Genetic engineering] is a reality with which we cannot fail to reckon," and "Now," says Zorza, "he has a party decree to prove it."

For all this, we must face the fact that in all these countries we are moving forward on a basis of ignorance, not yet of knowledge. The questions remain: Are the benefits great enough and the dangers small enough that we can proceed, given adequate safeguards? Are the risks to the wider public, and to the world, likely to be of such magnitude that the work should either be banned or performed only under conditions of total maximum security? Should everything be slowed down until we acquire the necessary information? On whom should the burden of proof lie?

According to Charles Hutt, the former director of the General Council of the Food and Drug Administration, the burden of proof rests with the scientists, whom he commended for their responsible attitudes. Nevertheless, it is up to them to show that no hazard exists. Probably most people outside the scientific community agree. Whether or not they do, it is essential that both those people within the scientific community and those outside who have legitimate concerns do not feel that they are being railroaded. Public participation is absolutely vital and the British habit of "closed" decision making is, I think, unfortunate. Charles Hutt also said: "I do not believe that the public's right should be affected by guidelines drawn up by any group which has not undergone the procedures for public participation laid down in the Administrative Procedure Act." To put it somewhat less ponderously and with another example: if the issue of whether or not to grant landing rights to the Concorde because of the possible environmental damage has raised so much hue and cry, then there should be no less public participation and debate over the technical applications of recombinant DNA, for the implications are no less serious.

Understandably, however, scientists are men in a hurry, and moreover, they feel, as Paul Berg emphasized, that "hurry" is a thoroughly relative term. They say they are not hurrying at the moment, just getting going. Before the advent of DNA methodology they were up against a stone wall, and they feel that too-strict guidelines will keep them there. Until the development of recombinant DNA technology the pace of scientific work in molecular biology had been infinitely slower than at any time in the previous twenty-four years. The pace of science is not a matter of whim or choice, Dr. Berg argues; it is a matter of what techniques and knowledge are available, and now scientists have a methodology that allows them to move forward. To those who say, "Slow down," Berg replies, "We have slowed down. We had decided to do this and we curbed the pace of advance." He feels that they have already compromised enough, and it would be unfair to ask them to compromise further.

Now that the guidelines are in operation, some of the people are sure to break the rules. How does one effectively police scientific activity? The very word "policing" is an anathema to scientists. But the fact remains that when a man goes into a laboratory he is taken on trust. There may be regulations; there may be a committee which *will* certify that the experiment *will* be conducted according to the guidelines laid down by NIH; there may be a biological safety officer attached to every institution, but the thought that he should actually *watch* the scientists at work raises adrenaline levels alarmingly. I persistently asked the scientists to consider whether if it were necessary they would be prepared to adopt the kind of procedure that is followed in coal mines in Britain, for example. Men are forbidden to carry matches down mines, and since so much is at risk, the unions have come to accept a set of procedures which thoroughly encroach on people's freedom and privacy. Before going into the cage, all miners are searched by one man, chosen at random, then he himself is searched by one man chosen at random. That scientists should accept unannounced spot checks on them or their procedures, or that safety officers should prowl around like a policeman on a beat, is regarded as outrageous and appalling. They all recognize that the problem exists, and if it *were* found as dangerous as coal extractions, would accept the control. But at least five distinguished scientists to whom I spoke said that people would only learn that this business was hazardous after an accident—and they still feel that such overseeing would be inimical to the progress of science, and is at present unnecessary. If the research is dangerous, then we can, according to them, perhaps expect a London School of Tropical Medicine or a Fort Detrick accident at some time. Then—and only then —will their sloppy colleagues be impressed. But, they argue, since we *accept* a certain cost-risk ratio in any hazardous industrial undertaking—so many injuries to so many men during so many years—the possibility of an accident should be perfectly acceptable. In any case they don't feel that the risks are at all as high as those we are willing to run every day in other areas of human activity.

Is there anything else we might reasonably insist on? Dr. Sydney Brenner says that he is a "firm believer in not trusting," so the best thing would be to install the discipline through machines, as with aircraft. He admits that one could certainly try to discipline by rules and regulation, by policemen or even by refusing to publish those papers that describe unsanctioned experiments and other forms of peer pressure. EMBO in Heidelberg has a high-risk facility with two policemen at the door, who see that people are not permitted to drift in, as I did elsewhere. Scientists may even have to be searched, though they may not be monitored in the laboratory. Brenner insists, however, and he is right, that all this is so destructive of the informal creative aspect of scientific research that it would be best to leave the whole thing to the machine—to put the whole procedure inside a box that can't be reached. Consequently, at Cambridge, England, they are trying to design a machine to do a great part of the recombinant DNA process within a completely sealed system, with remote control arms; if the finances are forthcoming, it is certainly well within the scope of scientific ingenuity.

In any case, sloppy people could be screened out. Dr. Kenneth Murray said he would like to choose very carefully those people who would be permitted to do this research and would like to have the ultimate sanction of firing them if they didn't conform. But though he feels he would have, or likes to think he has, the personal strength to do this, sloppy technique is not necessarily sufficient justification for firing someone in Britain today.

What is clear is that the situation *must* be under constant review, for as much as anything else it was complacency with existing procedures that led to the smallpox outbreak in London. Though industry is not controlled by NIH, it is clear that existing systems of law are sufficient to compensate the public after an accident (see Chapter 8). But compensation is *not* protection, and in both America and England, Health and Safety at Work Acts protect workers. It may well be that these acts will by themselves put a sufficient damper on the eagerness of industry to get into this area, or at least make certain they conform to the guidelines. Their eagerness should not be underestimated. One would like to think that legislation is not necessary, and that the remembrance of the thalidomide scandal, the recent chemical explosion at

Flixborough in Yorkshire, which almost wiped out a whole town, or the disasters at Hopewell, Virginia, and Seveso, Italy, would be sufficient to ensure that industrial concerns would be whiter than white, more careful than even the scientific laboratories. Brenner thinks they will be, but other people are not so sanguine; Senator Edward Kennedy is one of these. He thinks we may have to do more than just wait and see, for he agrees with the microbiologist who told Michael Rogers at Asilomar, "Nature does not need to be legislated, but playing God does."

We must therefore now turn to the political arena.

8
Creating New Precedents: The Law and Politics

We think that the true rule of law is, that the person who for his own purposes, brings on his lands and collects and keeps there anything that is likely to do mischief, if it escapes, must keep it in at his peril, and, if he does not do so, is prima facie answerable for all the damage which is the natural consequence of its escape.

—Judge Blackburn
Ruling in *Rylands v. Fletcher;*
Court of Exchequer Chamber, 1865

At the outset a distinction must be taken between animals mansuetae naturae, *and animals* ferae naturae. *The former class includes domesticated animals . . . which are not naturally dangerous. . . . Animals* ferae naturae *are those which are of an obviously dangerous or mischievous disposition, such as monkeys, elephants, bears, tigers, gorillas. If they do damage, their keeper is liable without the necessity of proving . . . savagery; . . . Nor was it any defence that the particular individuals of that class of animals are more or less tamed.*

—George Winfield. Textbook of
The Law of Tort. Cambridge, 1943

The story is told of a judge who presided in a case which turned on whether or not an animal was "naturally dangerous." In order to make the point clear the judge conjured up the vision of a lady who woke up one morning and found a tiger lying in the bed beside her. She immediately had a heart attack and died. Was the keeper of the tiger liable? The judge ruled that he was, for "it matters not how amiable the disposition of the tiger; it is dangerous by nature."

It is not entirely frivolous to ask whether these Brave New Bugs will be animals *ferae naturae*—that is, dangerous by nature. In the discussion of the wider implications of recombinant DNA research, the question has already arisen: shall we have to strain to come up with new legislation to cover the possible escape and damage caused by these new organisms? Not at all. As the scientists at Asilomar discovered, our existing system of common-law protections, evolved over the centuries, is quite adequate to give remedy in law to anybody who suffers damage. But, as we shall see, though society has legal remedies available against escape and damage, it presently has no legislative sanctions for effective monitoring or control in order to *prevent* escapes and damage.

It all goes back to *Rylands v. Fletcher* and Judge Blackburn's ruling in London in 1865, which has formed the basis for judgments in similar cases in both Britain and America ever since. The judgment is a perfect example of a law which both covers existing needs and has the necessary flexibility to cover new ones. The ruling states that if for your own purposes you collect any material on your land that might damage people if the material escaped, and if it does escape and cause damage, then you are liable—even though it was not your negligence that led to its escape in the first place. We might think this would apply only to things that escape on four legs or two wings, but in the original case, it was water, not animals, that caused the trouble. A farmer employed an independent contractor, whose competence was never at issue, to build a reservoir. While doing the work, the contractor came upon some old mine shafts that communicated with the mines of a neighbor. No one suspected this, for the communicating passages appeared to be filled with earth. When water filled the reservoir, however, it burst through the old shafts and flooded the neighbor's mines and land. The neighbor sued, and the original farmer was held liable—even though the flooding was the result of a series of events that he could not have foreseen.

The judgment did not rest on *anyone's* negligence. The farmer was responsible because no mischief would have occurred if he hadn't brought the water onto his property in the first place. And, Judge Blackburn went on to say, "This, we think, is established to be the law, whether the thing so brought be beasts, water or filth or stenches" . . . or recombinant DNA?

Most of the time people do not use this remedy because suing is expensive and the outcome doubtful. The late Justice Goddard—a famous English judge—has said, "Litigation is just like backing horses, but much less certain." Nevertheless, the *Rylands v. Fletcher* ruling advanced the existing law in two ways: one, by making the occupier-owner vicariously responsible for faults or actions of *other* people, including servants, independent contractors—anyone, actually, except a stranger—and two, by extending to the escape of things a liability that had been obvious at that point only in the form of fire or cattle or unruly beasts! Over the years, as more and more cases have come up, the rule has been applied to a remarkable variety of things: fire, gas, explosions, electricity, oil, noxious fumes, colliery spoil, rusty wire from a decayed fence, vibrations, poisonous vegetation, a flagpole, and a "chair-o-plane." There is no doubt that bugs from recombinant DNA may fit naturally into the category of "mischievous things," and it was a shock to the scientists at Asilomar when they realized this was so.

It would be unfair to say that the lawyers at the conference —Daniel Singer, Howard Green, Alexander Capron and Roger Dworkin—ran rings around the scientists, but they certainly ran rings around the objections the scientists raised to the legal implications of DNA research. When it came down to technicalities, these four lawyers were as competent and confident on *their* ground as the scientists had been earlier about the detailed manipulations of a short strand of DNA.

It was Roger Dworkin, professor of jurisprudence at Oxford, who best laid out the legal tools and remedies available to the general public. Dworkin felt the scientists had better know all about them; otherwise, they might find the issues sneaking up from behind in the form of multimillion-dollar lawsuits. His views are now summarized.

Remembering that the people who had actually been sued in the thalidomide case were representatives of the drug companies, not the actual scientists who discovered the drug, I suspect that the scientists, if they thought about it at all, had been relaxing behind what they took to be established precedent. As we have seen in Chapter 5, there has never been a contract between the scientist and the public, so the law of negligence and malpractice has never applied. The contract between a professional, whether lawyer, plumber or accountant, and a layman states that in return for a fee the professional must always exercise his skill to the best of his ability, and in his client's interest, not his own. The moment he falls short of his best skill or knowledge or ability, his client can sue. One could sue a scientist only if one had a specific contractual relationship with him or her.

Nevertheless, *if* as a result of the scientist's actions, recombinant DNA escaped and damage followed, then, as the Asilomar gathering discovered, the legal remedies could be very painful.

First of all, the legal programs already surrounding scientific research are applied by a wide range of decision makers, not many of whom are scientists. In America, most of them are to be found in the Office of Safety and Health. Britain is similarly covered by the Health and Safety at Work Act of 1974, where the principal authority rests not with the Department of Health, but with the Secretary of Labor. Similarly, tort procedures, whether in England or America, are administered by people whose only expertise is in the law, not in science. The major role in all legal cases is retained by the jury, by lay people who have been built in to the existing system *precisely because*—as Dworkin insisted—*they are experts in nothing.* It is no good for scientists to talk about having these issues assessed only by experts, for it is just not going to happen that way—as recent events have shown.

When injured people sue under the tort system, they do so in front of common-law judges and juries, who pay considerable attention not only to compensation for the victims, but to the conduct of the person who did the injury and to the control he exercised. In a hypothetical case involving recombinant DNA research, the decision as to the extent of liability would depend on whether or not the researcher had achieved a desirable level of control. And as *Rylands v. Fletcher* showed, one may be liable for accidents, even if there was no *intentional* injury or *intentional* negligence. In addition, people injured by DNA research would be able to recover not only from the people who did the research, but in all probability from the employers and the institutions who did the funding too.

Aside from the unforeseen and unintended accident, an injured person might raise the issue of direct negligence. Dworkin emphasized that this implied two kinds of obligation. A lawyer could argue that the scientist was negligent in the performance of the research and the safeguards provided by not following the right procedures. The second argument would be even more worrying, for it is also possible to contend that the scientist was negligent *in doing the research at all!*

Negligence is "unreasonable conduct" and is not quantifiable, so the questions that the judge or jury would want answered would be: How likely was the action to hurt someone? How bad was the injury? How important was the conduct that produced the injury? How much would the freedom of a man be infringed if we asked him to desist? How expensive would that be? These far-reaching questions which cover freedom of inquiry and the necessity to civilization of scientific research would once again be decided by lay people, Dworkin insisted.

What kinds of defense could one make about issues of "reasonableness" or "unreasonableness" in the actual performance of scientific work? A defending counsel could point to the NIH guidelines or other categories, to the way a scientist had complied to the customs of others, or to official or semiofficial codes of safe practice. This might make it more difficult to prove a scientist's negligence but would not make it impossible. Scientists following codes of practice laid down by NIH could *not* be guaranteed immunity against prosecution, nor be certain that the case would go in their favor. The scientists were reminded of a very famous case when a judge ruled that tugboats should be fitted with radios, for, as he said, there are times when an entire industry lags behind the standards of safety which must be imposed, and which society has a right to demand.

Nor could a scientist take refuge by appealing to questions of "agreed standard practice." In 1974, the Supreme Court ruled that an ophthalmologist was negligent for failing to perform a glaucoma test on a patient, even though the patient had no symptoms whatever and was under forty years of age, and despite the facts that it is standard practice *not* to do such a test at that age and that the likelihood of the person having glaucoma was only one in twenty-five thousand.

Scientists confronting the law are no different from anyone else. They might try to argue that they are only liable to those they have a duty to be non-negligent toward—i.e., their technicians, janitors, and so forth—but it will not help, because the court will hold that they have an obligation to anybody who might be injured, and that, he said, means *everybody.* Perhaps a bizarre situation—such as a burglar coming into the lab, taking the recombinant DNA and spilling it all over town—might take the scientists off the hook, but it would not guarantee it.

How about those people who, like janitors, choose to work in a laboratory? Are they not assuming the risk, legally? Not any longer. There was a time when, if you explained the risks very carefully and your technician took the job, the risks were up to him. But now the courts believe the job market is such that people no longer have the option to say, "To hell with you, I'm not going to take this job."

There was one other devastating category to consider: the category of "abnormally dangerous," for which there is no doubt about strict liability. What constitutes "abnormally dangerous" is always a matter of debate, and in court a list of factors have to be examined to decide the degree of danger. These cover the severity of the risk; the gravity of the harm; the fact that even by being thoroughly reasonable, the risk could not be eliminated no matter how hard one tried; the question of whether the conduct was common conduct; the appropriateness of the place where the work was going on; and the value of the activity to the community. In order to prove "abnormally dangerous conduct," however, not all the factors need to be present, and recombinant DNA research plainly qualifies as "abnormally dangerous conduct" on the first four counts. There is perhaps a difficulty over the fifth as well, for while it is possible to argue that any university is an appropriate place to do such research, it is, as Dworkin pointed out, equally possible to conclude that such activity should be restricted to isolated laboratories in the middle of the desert. As to the value of the activity to the community, this is totally unknown at the present. Dworkin warned the scientists that if they wanted to rely on this point in order to avoid a lawsuit, they had better be a great deal more specific about the value to the community than they had been so far.

Could a scientist get off the hook by saying, "I was told to do it"? If *ordered* to do it by the government for defense purposes, then probably yes, and the government might become liable. But federal funding per se is not the same thing as a government order. In any case, all employees have a right to wide safeguards and employers have a duty to provide safe places in which to work, *free from hazards:* not reasonably or partly free, but *free* of recognized serious hazards. Where toxic materials or harmful physiological agents are employed, the place of work must show that no employee will suffer material impairments. Therefore, the employer, in both Britain and America, must conform to the best possible standards based on the best available evidence.

Under workman compensation acts, an injured party needs to demonstrate very little: neither negligence nor abnormal danger, nor that the procedures were or were not safe. Anyone injured on or because of his job can and will be compensated. The sole issue is how much he is to be given. If the government ordered the scientist to do the work, then the government alone could be held liable. Otherwise both the government *and* the scientist get nailed. The Occupational Safety Health Act calls for direct government regulation of places of work. The penalties for infringement are payable to the government and are backed up by criminal sanctions. Dworkin turned the screw even tighter: these penalties do not replace other liabilities, they merely supplement them. In addition, if the Secretary of Labor is unsatisfied about the protection provided, he can insert emergency safe standards for a period of six months, pending an investigation, *without any kind of appeal.* This would require the people running such establishments to show a log of procedures. The secretary would be empowered to enter and inspect the premises without notice, and criminal penalties would exist for any person who might tip off a scientist before such a visit. Willful violators of the code of practice can be punished, first by citation and then by civil penalties that run from $10,000 and six months in jail to $20,000 and one year in jail!

The conclusion from all this? Scientists can be liable—devastatingly so—for their actions as they relate to society, and this, Dworkin insisted, made it all the more urgent that they invite public participation in their endeavor. Society provides the matrix for science, and scientists should not imagine that they can monopolize the policy role. "Academic freedom" does not encompass freedom to do harm to others, nor does the "scientifically justifiable" experiment. The lawyers emphasized at Asilomar that scientific interest in itself is *not* sufficient argument for proceeding with an experiment—though admittedly this is a value judgment with which people will disagree. Crucial as the work may seem to the scientist, the public may not concur, and in a democratic society, it is the public's right to come to the wrong decisions. Scientists may lobby; they may protest against such decisions—indeed, the lawyers said they probably have an obligation to protest—but they must abide by them.

In addition, scientists as a group just may not be competent to specify the risk-benefit ratio. They can suggest how likely the risks are, though, as they admitted at Asilomar, they mostly do that by hunch, but the degree of recognized benefits is a social decision, and it is dangerous for the scientist to pretend otherwise. When Dr. Joshua Lederberg asked about the hazards of unreasonable harassment from lawsuits that had no proper basis, the lawyers were not too sympathetic. They pointed out that when scientists seemed to be worried about the hazard of unreasonable harassment, what in fact they were asking was: would there be unreasonable restrictions on their work, with the law mucking up what they wanted to do? It was no good taking the stance of one scientist, who was heard to say, "All we have to do is to come up with a piece of paper which will provide reassurance to the public that we are proceeding in a reasonable manner." This, the lawyers said, smacked of cosmetics. The job of scientists was not to provide cosmetics. Their job was to protect the public from harm. If someone came to them and asked, "What can I do to protect myself?" a lawyer can only answer, "You will have to do what is the right thing. And you will have to make it a matter of your own judgment as to what *is* the right thing."

The burden of proof is now on them, not on society. Over and above all other questions, the law has traditionally placed a special responsibility on the learned professions. They have an obligation to pass on an awareness and a sensitivity to the ethical codes that supposedly govern their actions. Scientists have traditionally passed on standards of good scientific behavior to others, but now they are in a new situation. They and their apprentice scientists need to be made quite aware of the effect of their own behavior on others at risk. It is up to the proponents of the experiments to show, first, that the dangers are not trivial, and to specify exactly what they are, and second, that the benefits are so certain and overwhelming that it is reasonable to allow the procedures to continue. The lawyers at Asilomar doubted whether science was yet in that position. It is not the job of society to prove the experiments are dangerous. Scientists must convince people that the social benefits being promised are so good that everyone—society and scientists—should be conscripted to assist in the undertaking. In the last analysis, they emphasized, the scientific profession must now face the possibility that some experiments in genetic engineering might be scientifically elegant and socially beneficial but morally reprehensible.

The lawyers' message was not a soothing one, but then it was not meant to be. With few exceptions, of whom Sydney Brenner was undoubtedly one, the lawyers understood the political mechanisms surrounding the profession much better than the scientists themselves. It was not that they were negative. They emphasized what was true: that as experts, scientists command a great deal of power and a deep reservoir of regard. Many people are still mesmerized by science. If, as a profession, scientists bring the public along with them and show that they exercise reasonable control, they will probably continue to receive the highest respect and recognition. However, there is a history of disaster for those professions who have not used their opportunities wisely— medical practitioners, for instance.

Initially the law favored physicians, and it was almost impossible to find two professors of medicine to give evidence against a practicing doctor. Consequently, for years it was difficult for an ordinary person to recover judgment in a malpractice suit. Physicians abused their trust, and thus through a combination of factors—an aggrieved society, greedy, percentage-fee lawyers, and physicians' one-time refusal to cooperate with the system of sanctions—medical practitioners are now being massacred in the courts.

Thus any suggestion at Asilomar, or later, that scientists were acting in what appeared to be their own self-interests would rebound on the profession. If scientists were worried about their own freedom, if they didn't like saying to each other, "You can't do that experiment"; if they were unwilling to develop adequate procedures for overseeing whether their colleagues were meticulously careful in the laboratories; if in fact they did not come up with adequate regulations stricter than those they really wanted, then the odds were that the law would impose much more rigorous ones.

After this session at Asilomar, no scientist could be in any doubt that the issues had many more dimensions than the strictly technical ones they had been discussing. Yet those among them who from the very beginning had been acutely aware of the political implications were disillusioned as well. The lawyers continually pressed home the point that the scientists were *not* in a position to make policy. So it was a shock when, two months after Asilomar, Senator Edward Kennedy claimed that was precisely what they had been doing. As he told an audience at the Harvard School of Public Health in May 1975, "It was commendable that scientists attempted to think through the social consequences of their work. It was commendable, but it was inadequate. It was inadequate because scientists alone had decided to impose the moratorium, and scientists alone had decided to lift it. Yet the factors under consideration extend far beyond their technical competence. In fact they were making public policy and they were making it in private." Kennedy's statement bewildered, stunned and even angered members of the community who had taken for granted that the Kennedys were always on the side of the angels. Ever since the mag-

nificent impetus given to science by the late John F. Kennedy
and his scientific advisers in and around the White House,
led by that energetic figure, Dr. Jerome Wiesner of the Massachusetts Institute of Technology, scientists had regarded
the Kennedys as certain protectors of the enterprise and its
ethos, as a firm bulwark against the encroachments of politicians, bureaucrats and general snipers. So when Edward
Kennedy made it plain that he, for one, was no longer prepared to accept cherished assumptions and hallowed attitudes, there was a sense of personal betrayal. If hell hath no
fury like a woman scorned, this is equally true of a professional group who see one of their staunchest defenders apparently turning on them.

In 1971, Senator Kennedy took over the chairmanship of
the Subcommittee on Health of the U.S. Senate Committee of
Labor and Public Welfare. Since then, his own political nous
and the energetic investigations of his staff have disclosed
attitudes they have found arrogant and assumptions that
they question. By now a degree of mutual distrust has built
up between him and certain members of the profession, none
of which deflects the senator from what he considers to be
the proper course in this matter.

Kennedy has challenged a number of assumptions and
offered several pungent criticisms. One assumption is that
scientists know best how to go about their own business; the
challenge is that the scientific community fails to respond to
the public's needs, that even if the scientists are aware of
them, they ignore them. Arguing on the well-known principle that he who pays the piper is entitled to call the tune,
Kennedy wants far greater public participation for the
American taxpayer, who should, he says, be able to influence
not only what research is done, but when, how and in what
form.

His legislative activities have had some very concrete
practical results in research areas. He and his staff created
legislation that led first to a moratorium on research on human fetuses and then finally to a National Commission for
the Protection of Human Subjects of Biomedical and Behav-
ioral Research. (Another progenitor of the commission was
then-Senator Walter Mondale.) By law the majority of the
members of this commission are nonscientists. Kennedy is

also considering a bill that would set up a similar commission to review *all* basic research, even that which may *not* involve human subjects. The work on recombinant DNA—in fact, everything to do with what is commonly known as genetic engineering—would come under such a commission. In that May 1975 address to the Harvard School of Public Health, Kennedy reminded his listeners that when he undertook to make a thorough examination of the programs and policies of the National Institutes of Health, he received a "considerable degree of support from the public." He noted, too, that

> Academia has been on the defensive. It has chosen to view public scrutiny as a threat to scientific independence. It has chosen to view public involvement in particular research areas as inappropriate, and representative of a trend towards anti-intellectualism. . . . "I believe this élitist and acutely parochial approach does not serve the country well. It is an approach that will ensure continuing tension between the scientific community and the public. . . . The tension arises from the concern about how research priorities are set up by the scientific community. It arises from a concern as to whether certain research should be done at all, and if so, under what conditions. Finally it arises from a concern about the safety of some research prospects.

His concern about recombinant DNA research relates, on the one hand, to the public's input, and, on the other, to the political control that Congress might choose or be forced to exercise. By now his subcommittee has held two hearings on this matter. The first, on April 22, 1975, was pointedly entitled "Examination of the Relationship of a Free Society and Its Scientific Community," and in his first address Kennedy specifically stated the questions at issue:

> What is the nature of this research which so disturbed the investigators that they felt compelled to stop it for a time?
> Is there a safety threat to the general population?
> What are the implications of the research for the society as a whole?
> Was it proper for scientists alone to decide to stop and then resume the research?
> How could non-scientists participate in the process? Even if that were desirable, what should be done now in terms of public policy in this area?

> What are the potential dangers of Federal intervention?
> The issues being raised stretch beyond this form of re-
> search. They go to the heart of the relationship between a free
> society and its scientific community.

The arguments at the hearings, many of them familiar to
us by now, raged back and forth between supporters of more
controls and those opposing them. Of the scientists who testified, some felt that public participation was welcome—but
only in a limited form. Dr. Stanley Cohen suggested that the
handling of radioactive isotopes presented a good model of
public participation. Radioactive materials are subject to regulations designed to protect scientists and society alike, and
the public has channels by which they can be involved in
the enforcement of those regulations. However, the merit or
lack of it of specific experiments that involve isotopes remains a scientific judgment entirely, subject only to peer
review. The public does not require scientists to justify this
experimental tool in terms of the social benefits the experiments may bring. In the same way, Dr. Cohen believes, provided the public were assured of the safety of the procedures
and had adequate means of monitoring the safeguards, "it
would be contrary to the public interest if the initiative of
the scientific community in raising issues of experimental
safety should lead to a decision by the public to direct the
course of such investigations."

Similarly, Dr. Donald Brown of the Carnegie Institute insisted that scientists have the special knowledge to recognize
the potential hazards of their work and devise constructive
solutions, and that the participation of the public should
come in only at the point of "the practical application of
scientific discoveries and their moral and ethical consequences."

There were others, however, with very different views:
Dr. Willard Gaylin, president of the Institute of Society, Ethics and the Life Sciences, and Dr. Halsted Holman of Stanford
University. Indeed, Holman went so far as to argue that the
model of informed consent which presently applies to human experimentation in medical research should be the
valid one for all basic research in science. The right to experiment is neither divine nor god-given, he said. It was not a

question of assenting to dangerous experiments, but of assenting to *any* experiments. He suggested that a national
commission be set up to oversee science, and a few weeks
later recommended that all universities be required to hold
public hearings about research on campus. Several have
eventually taken place, including two important series of
meetings at the University of Michigan and in Cambridge,
Massachusetts (see Chapter 9).

Probably the most important point philosophically was
made by Dr. Gaylin. Genetic engineering, he said, is merely
a prototype of a group of problems which represent certain
common conflicts between the scientific community and the
rest of society. What is new in the history of the profession
and society is the occurrence of episodes that the public
construe as the failures of science: the disastrous effects of
the atom bomb or the gross perturbations in the natural
cycles discovered by ecologists. These episodes shake the
public's confidence in science as the bearer of good gifts and
highlight the Janus-faced nature of much of twentieth-century scientific and technological progress. Not only do these
problems exist but their very existence is due to the successes of science. As Gaylin put it: "It is the very success of
technology that has redefined its pursuits as public concern."
Where the issues are now life and death or the modification
of human behavior and the alteration of the nature of man,
science has transcended its original purpose, which was the
understanding of nature, and in the process its issues have
ceased to be wholly scientific. They have become moral and
political as well.

Kennedy was not satisfied with what he learned at the
hearings. Much of the scientists' testimony seemed entirely
too self-protecting and too inclined to ignore the larger issues. Equally, the hearings themselves had not been well
attended: only one other member of the subcommittee appeared—and he only stayed fifteen minutes. It was plain that
another hearing had to be held, and on September 22, 1976,
Kennedy convened it, this time assuming that the public had
the *right* to regulate scientific pursuits, and focusing on the
specific questions: does the public have the knowledge, expertise and ability to regulate science, and, if so, what are the

most appropriate methods for doing so? (This hearing was marginally better attended: three senators were present, out of a possible fifteen, but two stayed for only a short period. That is par for the course, I was informed.) In his opening remarks, Kennedy emphasized the important precedent that was being set:

> It is that the implications of technological advances must be carefully considered early on, and must be considered in public processes with wide participation from as many diverse elements of society as possible. . . . The issues go far beyond safety questions—in many ways, those are the easiest to answer. The real problem is to understand the social consequences of what science can now enable us to do. . . . I believe the debate over genetic engineering must go on. Scientists must tell us what they are capable of doing, but we as members of society must decide how it should be or whether it should be applied. Congress cannot legislate an appropriate answer in this matter. But it can and should take the lead in assuring that these issues are discussed publicly and by as broad a segment of the population as possible.

That said, the hearing got down to the specifics of sanctions and control. One glaring fault—and need for legislation —immediately became apparent. The committee revealed a great potential for abuse of the guidelines: many groups doing research did not and *still* do not fall under the guidelines at all. Only research under grants from the NIH and the National Science Foundation was controlled. Other agencies such as the Departments of Defense and Agriculture and the CIA were excluded, but so was industry—the source of a potentially enormous amount of laboratory experimentation. Senators Kennedy and Javits believed closing such loopholes to be a matter of great urgency and had sent a letter to President Ford in July 1976 urging him to remedy the situation by presidential fiat if necessary. At that point they had received no reply—it finally came in the day of the hearings. By early November an interagency commission had been set up to look at the problem so far as the federal agencies were concerned. That apart, the senators were extremely skeptical both about the industries' willingness to conform to the guidelines without a great deal of protest, and about our capacity to monitor them—or anyone else, for that matter—

if they did become subject to the regulations. This skepticism was heightened, and the concern stimulated into forceful reaction, by the fact that the General Electric Company did not testify at the hearings. In response to the invitation, GE had expressed its regrets but had said that its relevant staff member was abroad. So in a second letter, GE was asked to be kind enough to send along that staff member's deputy. GE replied that this person also had other commitments that day. Kennedy's staff drew their own conclusions. In addition, although the Pharmaceutical Manufacturers Association attended, the Manufacturing Chemists Association turned down the invitation issued by Kennedy's staff, with the terse comment that "they were not authorized to speak."

Such coy evasions may well rebound, for Kennedy said: "The actions of General Electric made it clear that we will have to have a complete accounting of all industrial research and development in this area."

All the witnesses, without exception, agreed that the guidelines were exact, tough and, if adhered to, adequate, but that they should apply universally and that all research activity should be registered. (*Both* these conditions now apply in Britain, under the aegis of the Health and Safety Commission, a responsibility given to it, it must be said, over the heads of protesting British scientists and the civil servants in the Department of Education and Science.) The problem of monitoring and control, however, is another matter everywhere. What was revealing and somewhat disturbing at the hearing was the extent of agreement about the ineffectiveness of such controls, based on past experience. Under the pressure of questioning, even the agencies themselves admitted this ineffectiveness to be a matter of concern. It was all very well for Dr. Donald Fredrickson, the director of the National Institutes of Health, to make reassuring noises, but Kennedy was able to point to the "weakness" of the FDA in monitoring scientific information that had been supplied by some of the independent research groups, and to similar difficulties encountered by the Environmental Protection Agency and to yet other ones, revealed by a report of the Senate Subcommittee on Administrative Practices and Procedures, regarding the National Cancer Institute. In those hearings, Dr. Frank J. Rauscher, Jr., the director of the National Cancer Institute, had admitted that there had been insufficient monitoring of the safety and effectiveness of the testing of some five hundred chemical compounds for cancer-causing properties. These oversights ranged from safety conditions for the workers to questionable contractual arrangements. So slack had the coverage been that the subcommittee had recommended some drastic measures, including "unannounced and more frequent and thorough site visits to the bioassay testing laboratories."

With all this as background, Senator Kennedy's question to Dr. Fredrickson about recombinant DNA was right on target: "Why is the public going to believe that it [monitoring the record] is going to be very much better in this area?" Though Dr. Fredrickson tried hard, it was clearly impossible for him to give any reassurances that the situation would be better in the future.

In conversation with me, Dr. Sydney Brenner said he felt that industry would try very hard to be whiter than white because they could not afford a mistake here. Yet three months later (December 25, 1976), *The Economist* reported that seventeen drug and chemical companies in the United States asked for exemption from the government rules on genetic engineering on the grounds that submitting research plans for approval would affect the possibility of getting patents. This *may* only mean that the companies would still follow the recommended safety procedures, and merely keep the *details* of the experiments under their hats. But it may not, and *The Economist* justifiably concluded: "Genetic engineering should be taken more seriously than that."

Indeed the whole problem of patents could be something of a can of worms. A number of universities are currently reviewing patent applications for discoveries in DNA research, and Stanford and the University of California have already filed claims. Industry will surely follow suit as their research projects expand. The problem is that once the ownership of the patent is given, the inventor can subcontract the development, and though the initial research might be conducted under the guidelines, no such obligation presses on the developer. NIH is being forced to consider permitting inventors to hold patents only on the condition that any licensees must also follow the guidelines. Unfortunately, they cannot simply declare it. With the granting of a patent, control passes from public to private hands, so this is another area which will certainly eventually be covered by legislation.

One way or another, the issue of industry conforming or not conforming to the guidelines is a delicate and pressing one, and it would seem that a degree of skepticism is not unjustified. One week before Gerald Ford left office, and just before Jimmy Carter's inauguration on January 20, a sharp-eyed member of Senator Kennedy's staff was scanning the *Federal Register.* This is issued daily by the Department of Commerce and lists the regulations and new rulings as they develop. On this day the register contained the startling information, quietly inserted, that all patents relating to recombinant DNA research and its applications could be expedited. All patent applications have to take their place in the queue and processing them can take some time, but this ruling moved these patents way up to the front of the line. There was one caveat: the work patented must follow the NIH guidelines. But this was followed by another caveat that could well neutralize the effect of the first, for it said that deviations would be allowed if the circumstances warranted. Kennedy's staff are curious to know what pressures were applied, and by whom, on the then Secretary of Commerce to expedite this matter. In any case, they wrote to the new secretary, Juanita Kreps, asking her to reexamine this ruling in the interests of the public. They were justifiably annoyed that the ruling had been made without discussion and without any opportunity for public participation in the decision.

The whole episode leaves a sour taste: the need for some legislation to cover both industrial work and sanctions against those who transgress the guidelines would seem to be paramount, but at the same time we must make certain that the opportunity exists to revise the guidelines, upwards or downwards, as our knowledge develops. Though there is nothing in present legislation that imposes the guidelines on industry, there is strong moral pressure. Just what impact this will have remains to be seen. In any case, they are at the moment restricted by certain bans, which prohibit release into the environment of materials or organisms made by these techniques. To say that industry is unhappy with this state of affairs is an understatement; they will certainly try to get these restrictions lifted by all means possible. The moment they have made a valuable product, like human insulin, and can see the possibility of full-scale commercial application, they will probably try and get a test case through the courts. So the law may come back to this problem by yet another route.

Some other very practical and concrete suggestions emerged from this hearing, over and above the general insistence that some mechanisms must be found to extend the scope of the guidelines to all recombinant DNA research in the United States. Dr. Norton Zinder of the Rockefeller University felt that since "some pretty bad scenarios can be written about this technology," but since no one knows for sure whether those scenarios are real, it would be worth doing some experiments under appropriately safe conditions to see whether recombined organisms that might be dangerous could exist and survive in nature. Dr. Robert Sinsheimer felt that the risk of a "biological chaos" was great, and that some of the projected benefits of the recombinant DNA research could be gained by some less dramatic and potentially less harmful means. He endorsed Dr. Halsted Holman's suggestion that the mandate of the National Commission on the Protection of Human Subjects be extended to give it authority over this field of research. This is now likely. Dr. Holman also suggested that NIH expand the guidelines in the area of epidemiology, and that there should be constant review mechanisms as the work proceeds and understanding grows —a suggestion that everyone endorsed. And he believed that all patents in this area should be denied.

It is likely that as a result of Kennedy's energy and sense of urgency many of these recommendations will ultimately be put into effect. What new steps he will take remain to be seen. Certainly, new hearings are planned. While it is clear that no new laws are needed so far as legal redress is concerned for damage from recombinant DNA research, new legislation is certainly needed to cover the loopholes and protect society fully. Legislation being at best an unwieldly instrument, the new laws should be small in number—but tightly applied and enforced.

It may well be that this present insistence on public scrutiny and public participation will by itself solve the problem
of the directions that scientific research and genetic engineering will take, perhaps even the form that sanctions will
ultimately take. Social and moral disapproval are not always
easy to withstand, even though Russell Train of the Environmental Protection Agency is surely right when, giving testimony at another Senate committee, he said: "We should
not be under any illusion that legislative acts will protect
us from willful action." Too restrictive legislation could
possibly be useless, even counterproductive, as far as genetic engineering is concerned. Scientists can always go to
Switzerland, where from Geneva *The Economist* reports
that dangerous experiments are already being done, or do the
"Saturday-night experiment." But the pressure of public
opinion may in the long run turn out to be forceful enough.
What we are seeing now is, I suspect, only the beginning of
a continuing trend.

In any case, the wish on the part of a few scientists to
regain full control of this research is now totally in vain. If
anyone is in any doubt about this, they need only read Senator Kennedy's summing up of the hearing, in which his
intentions were made plain.

> I think we have seen as a result of the hearings that there is
> an enormous area of uncertainty surrounding this whole area
> of recombinant DNA research.
>   No one can predict with any degree of absolute certainty . . .
> what the dangers would be, or what the opportunities and
> advantages would be . . . When some of our most skilled . . .
> and thoughtful scientists and researchers point out the dan-
> gers, they spell out a real living holocaust for this nation . . .
> and the world, and when they use their minds to elaborate on
> the possibilities of benefits to mankind, [they show that these
> benefits] are virtually unlimited.
>   We realize these guidelines should in no way be thought
> to be fixed to concrete. They should be adjusted, changed,
> altered, as the situation requires. They should be strength-
> ened [as] new dangers are presented. They should be lifted as
> the concerns for safety are eased. Therefore, it is extremely
> important that we have a continuing and ongoing review of
> these guidelines. . . . [and] we have seen from our recent past
> the importance of compliance with them.

The record of various governmental agencies during this history, and the work that this committee has done in a number of areas, has not been reassuring. Therefore, this committee is going to assume a very significant part of its responsibility insuring that those that do have responsibility in government will insure compliance with those regulations, and we are going to monitor that very, very closely in the future.

We are encouraged by the administration's willingness to insure that the Department of Defense comply. We are troubled by the apparent failure of the sense of urgency by the Executive in assuring that those NIH regulations will be applicable to all governmental contracting and research. I think it is unjustified and inexcusable that it has not done so. . . . We are distressed by the fact that some—associations, organizations, and industry—have not focused on this issue, which is a charitable way of viewing their response to our inquiries. Others have clearly demonstrated lack of [the] concern which I think is absolutely essential if we are going to really provide the types of protections needed for the American people. We are going to press the various associations and industry to assure that their protections comply with these guidelines.

Finally, I think it is absolutely essential that the public be brought into this process. I think if there is a continuing echo from the course of these hearings, it is the importance of the public being informed about this particular issue, understanding the potential hazards and benefits. I believe that they are fully competent to try and sift through the complexities of the potential dangers. If it is laid out to them in a positive presentation, I think that this *will* really ease the dangers of demogoguery on the issue—the scare and frightening tactics which have been the case in some other extremely important issues of public concern.

Yet even after all this some questions remain, those Kennedy raised at the very beginning.

The plain fact is that genetic engineering has the capacity to change our society. How do we want it changed? What uses can we make of this knowledge? What degree of change is desirable, and at what rate? What kind of society do we want to become?

But what no one mentioned is that some members of society may not want society changed at all, at least not in the directions that seem likely to stem from this research. These

people do not want this research. They are more concerned
with the quiet consolidation and preservation of what we
have and what we know than with the restless striving for
perpetual change, modification, new knowledge and its application. As we shall see in the next chapter, they argue that
there is no crisis in man's condition that calls for such a
hurried extension of scientific knowledge and its application.
If there is a crisis, it is of a totally different nature: it is moral
and spiritual, for which science has no remedy. Once again,
these concerns surfaced when the problem of recombinant
DNA research arrived in the wider arena of the total society
and the university communities at large.

9
Creating
New Moralities

Bones: We are not in ancient bloody classical Greece.
George: I absolutely agree with you.

—Tom Stoppard
Jumpers

In the years since the Asilomar Conference, the rip-
ples of consequence have spread ever wider and by
now affect much more than the technical concerns of a professional group. The question of experimental safety has en-
larged into a question of whether to do the research at all,
and technical issues internal to science have been engulfed
by ethical questions traditionally thought outside the profession: Is genetic engineering moral? Are there limits to the
freedom of inquiry? Is freedom of inquiry a right, or a privilege that can be withdrawn? Discussions once confined to the
profession have widened to include not only science administrators, lawyers and politicians, but members of the lay
public.

The advent of recombinant DNA did not precipitate these
discussions about the ethics of biology and medical practice.
The last few years have seen a mushrooming of debate about
science, values and morality. Institutions such as the Has-

tings Institute of Society, Ethics and the Life Sciences and the Kennedy Institute for Bioethics in America, and the Council for Society and Science in London are but three examples of "watchdog" organizations that have arisen, in response to both the problems emerging at the interface between science and society, and the social unrest of the sixties. By now, these institutions and the studies they encompass have had a considerable impact on universities and more importantly on public opinion and legislative bodies. If one has any worries at all, it is not about these institutes but about their university counterparts: how permanent will the interest be? Medical ethics is but the latest fashionable subject on campus, and philosophers long used to seeing funds for the humanities as only a minute fraction of those available for science are flocking to the fountainheads of newly financed programs. That many of them have never set foot in a hospital or talked to a dying patient and don't intend to do so seems not to matter. However, bioethics with its attendant workshops, readers, anthologies, courses, slogans, prophets and academic articles written for other medical ethicists may only turn out to be the latest hot subject, in a sequence that has included sex, women's rights and black consciousness. It may shoot skywards for a while, sparking as it flies, but finally fizz out.

Nevertheless, the issues remain, and during the fall of 1975 and the spring of 1976, the first full-scale public discussion of the problems raised by recombinant DNA took place at the University of Michigan at Ann Arbor. It was there that scientists, faculty and members of the community all came together for the first time, and during the stimulating days I spent on campus, I found reflected the whole range of public opinion on these questions. The debate that took place there may well prove to be a model for the kind of public participation in the progress of science that Senator Kennedy is hoping for in the future. As one faculty member said, it may well have been "a new social invention for deciding *who* decides, about what sort of research may be permitted."

When on May 11, 1976, the regents of the University of Michigan called a special meeting to decide whether the university should go ahead with the construction of a $302,000 facility to permit moderate-risk experiments in recombinant DNA research, much more was at stake than a building. If the university *as a university* made the decision not to encourage this research on their campus, the effect would be seismic, not only on local individuals like Dr. David Jackson, but on universities around the world. This was because scientists wanted at least one school to make a conscious decision on the work, to make it thoroughly "respectable" by giving it an unqualified blessing. Thus if after deliberation the school decided to halt further progress in this area, the repercussions would be widespread.

Six months before, the regents had been most favorably predisposed toward the project. Universities exist to push back the frontiers of knowledge, but this university's status in the biological sciences had been slipping. Recombinant DNA was becoming a hot field, and the presence on campus of Drs. David Jackson and Robert Helling presented the occasion for a full-scale push toward reestablishing preeminence in the field.

The ball started rolling when Dr. Charles Overberger, the vice-president for research sent a report to the Biomedical Research Council, under the chairmanship of John Gronvall, the dean of the University Medical School, urging that every effort be made to support the development and utilization of this "revolutionary technology," beginning immediately with providing support for the modification of three existing laboratories. The council concurred and went even further, calling for a major effort on the part of the administration to improve their position by recruiting additional faculty as quickly as possible and providing them with the best possible facilities. The interim solution of three modified laboratories pending the construction of a larger facility was simply not enough, they said. The Office for Research should regard the problem with a sense of urgency and seek other funds beyond those the university itself might be prepared to provide.

That did the job. The university formally agreed to give the researchers some $302,000 for the modification of the laboratories. On Thursday, November 20, 1975, the regents rubber-stamped that commitment. Even if the federal agencies failed to come up with the money for the project, the university would guarantee the amount. There were no conditions in this agreement, nothing discretionary, no contingencies. Everything was very low-key. The official line, as reflected in public statements, stated that the university would be doing low-risk experiments—even though, by Asilomar standards, they were already committing themselves to medium-risk experiments.

It was at that point that the biologists decided there should be some committees to advise in the development of the programs. Three were formed. Committees A and C were technical: A, composed of individuals who were either working with recombinant DNA or tumor viruses or were planning to, was charged with directing the modification of the laboratories and seeing what other major facilities might be needed. C was the committee of scientists called for in the NIH guidelines, a committee which would review proposed experiments and certify to the granting agencies whether the investigator had the facilities available to enable him to do the work in compliance with the requirements. No more than three members of that committee would work in laboratories that required such certification, and staff support would involve legal counsel.

Committee B, however, was the brainchild of Fred Neidhart, David Jackson and Robert Helling—and had most unexpected consequences for its creators. It was composed of individuals who shared an interest in biological research; the majority of them, however, were *not* scientists. They were drawn from various parts of the university—law, social work, English, philosophy, theology. With one other exception, the others were scientists who, though they did not work in the sensitive area, knew enough about molecular biology, I was told, to be able to "help the others along in spots where terminology becomes a little tricky." Their charge was to step back from the mundane problems of facilities and funds and look into the future of molecular genetics research, to develop and recommend a policy and/or review process concerning the social, ethical and legal consequences of this research and so advise the vice-president. It was, in many respects, a unique committee with a novel charge.

It was not meant to have a policing function. It could not unilaterally decide what kind of work should go on. But it could examine whether there should be formal mechanisms to ensure a dialogue between the scientist, who might be blinded by his own vision, and the humanist, the artist or the social scientist. These people, not having the same driving curiosity as the scientist, could from time to time raise the questions "What are you doing? What is turning up? Where are you going?"

It was a most laudable conception but it was quite inconceivable that its mandate should ever include anything that might stop the momentum. At one stage, the committee asked Dr. Neidhart the true devil's advocate question: "Can you imagine anything coming out of our deliberations which would be of any practical relevance to what is actually going to *happen* to the research at the University of Michigan?" The moment, I was told, was a rather embarrassing one. Though he was gently polite, as he always is, Dr. Neidhart in essence said that as honest, impelled scientists, they would go ahead with the research, no matter what the committee said.

In addition, though the majority of committee members were nonscientists, they were people guaranteed to understand the aims and motivations of the people whose work they were to examine. When I asked David Jackson and Fred Neidhart how they came to choose the members, they were disarmingly frank. They sat down over a drink one day with Robert Helling and made a list of those people whom they thought would be helpful, where "helpful" meant those who would understand. Although the committee might have criticisms, which would be genuinely welcomed, they would be given by people who were basically sympathetic, the three figured. They probably saw no reason to choose a committee any other way, and I have to emphasize that there is nothing Machiavellian about this procedure. This essentially "chummy" manner is the way in which most university committees are constituted, and is a usual aspect of academic politics. Other than a sympathetic approach and some suggestions, however, they probably never thought deeply about quite what else they expected from the committee's review.

However, at least one elected faculty group exists at the University of Michigan, which tries to look after the total interests of the faculty. In 1975, Professor Shaw Livermore, an historian, was a member. This group heard indirectly and unofficially about the creation of Committee B, and asked if one of their representatives could serve. The answer naturally was "Yes—why not?"—and Dr. Shaw Livermore became a member, the only person who represented anybody at all!

I visited Ann Arbor at the suggestion of Dean Gronvall, dean of the Medical School, who invited me to come "and see what was going on" when he heard that I was writing a book on this subject. On November 19, Vice-President Frank Rhodes, whom I have known for many years, arranged a small dinner as an occasion to meet some of Committee B informally: Rosemary Sarri of the School of Social Work, Carl Cohen of philosophy, and Shaw Livermore. That night, and in the days that followed, it became clear that some people on Committee B held a far different image of the committee —and of recombinant DNA—than the one the biologists held. Several things were obvious. The first and most forceful was a concern that Committee B would not be a mere rubber stamp, serving a cosmetic function *vis-à-vis* the research—making it clear that it had not been created simply in order to allay people's anxieties in the community, so that if someone said, "You scientists are fiddling about with life," the response could be "Oh, well, we have a committee." It seemed vital to some, both for their own intellectual integrity and for the public, to make it clear that they were not engaged simply in a laying-on-of-hands operation. At that stage, however, no one was sure.

There was a corollary to this: if the committee was not merely to have a cosmetic function, its members had to face the possibility that they might have to say no to their academic colleagues who wished to pursue a line of research— and not only that they might have to say no, but that in a very real sense they would be *right* to do so. This flew in the face of all academic tradition.

During the fall of 1975, it was questionable whether the committee felt this to be a problem at all, let alone that they had that degree of option, but Shaw Livermore, for one, was extremely exercised by the problem. While it was quite likely that he could in the end honorably give the go-ahead to the research, he had to have it fully understood that this positive decision could be made only if he felt he had a mandate to say no, if necessary. Otherwise there would be no option, and as he said, "I would feel unclean. . . . I would feel we were being used."

He was in fact raising questions about the very core of the existence of a university—namely, freedom of inquiry—yet he knew that Committee B would be *predisposed* to let the work go on, to find reasons for welcoming it, rather than to find reasons for delaying it.

Shaw had other worries on his mind as well, and so did I: moral disquiets, irrational forebodings, even an aesthetic repugnance for certain implications *arising* from the research. That night and in the days to come, Shaw and I began to articulate these.

One chief cause of concern is simple enough. It is not the short-term science questions—the safety measures already proposed and taken seem reasonably satisfactory. Nor is it an objection to motive. To want to understand how genes function is eminently reasonable. Suppose this technology is very successful, however, and man *is* provided with a new capacity for human genetic engineering; what would this entail? Dr. David Baltimore has given one answer: a profound qualitative change in our life. But why are we worried anyway? Surely it represents a typically science-fiction scenario which has been amplified over and over, ever since Mary Shelley. One tends to laugh it off as good reading, but no more. Perhaps, we concluded, our disquiet has roots not so much in the strictly scientific outcome as in the broader implications: a society manipulated on a massive scale.

We found this idea totally repugnant—but it also raises interesting counterquestions that should be examined first. The trouble is that one can so easily be forced into the impossible position of seeming to be opposed to all ways in which man has manipulated his environment. How does one distinguish between acceptable forms of environmental manipulation, genetic or otherwise, and unacceptable ones? Why are we prepared to countenance some and not others? While I was researching this book, scientists rightly threw these questions at me, and Dr. Fred Neidhart articulated their concerns most completely.

It is in the nature of man to achieve, to accomplish and to create, he argued, in science as in anything else. He could not see any especially new hazards arising from what biological scientists were doing. He suggested I turn to neighboring disciplines and look at the manipulations being carried out by psychologists or social scientists, or anyone working with behavior modification. There I would find techniques both more imminent and more directly applicable to human modification. Moreover, those same social scientists and neurosurgeons are cheerfully doing things which they do not thoroughly understand, he said. At this stage to tie recombinant DNA technology to a problem of potential human manipulation was unfair.

Nevertheless, the prospects are still there, the disquiets are still there. There is the "human-scare" scenario, by now familiar: evil people planning to use the technology to produce a race of acquiescent robots. There are the men of science so wrapped up in their methods that they can see no other way to proceed, such as the doctors who widely give drugs to hyperactive children and the two brain surgeons in Britain who recommended a brain operation to a psychotic's wife to help *her* cope. These are appalling dangers, but they are manageable. Our real concern was with the more likely problem, that of reasonable people who with great good will present to mankind a capacity to do things with mankind they may not find at all attractive.

That is the next problem: what does that phrase "Will not find attractive" mean? Does this mean that man seems to do well only when there are large uncertainties? On the one hand, scientists are perhaps offering us the possibility of eliminating pain, suffering, old age and aggressive behavior, and who could be against that? Yet on the other hand, one could argue that as we eliminate pain and suffering we eliminate feeling; that we would be escaping from the human condition and from human predicaments—those factors which have tempered and molded humanity throughout history, the expression of which has often resulted in great literature, music and painting. Again one can easily be driven into an impossible position. Certainly we would never say that it is man's fate to live in sorrow for most of the time, nor that suffering is good for you. Yet to attempt to escape from all human ailments and predicaments is probably disastrous; a tension-free materialistic Utopia would be fatal. Indeed, a number of scientists—Professor René Dubos is one—say that there are good biological reasons for stress —that man actually functions better under stress, for that is what he has become adapted to throughout evolution. Other scientists, however, will retort that if they are very careful, they can produce a human life from which great suffering is eliminated, and how could anyone be so callous or indifferent as not to want that? Thus every time one wants to express such disquiets, one finds that in their extreme form, they are very unattractive.

Similarly, no one wants to be in the position of saying that man cannot be trusted to cope with his capacity to manipulate the environment. We do this every day, every minute. And it would obviously be impossible to revert to a hunting and harvesting mode of existence, the last form of human and social life in which man did not consume or seriously disturb the environment.

The problem is that it is so hard to produce a *rational* argument for one's moral qualms about DNA research. To convince men of conviction and intellectual strength requires clean and persuasive arguments, and so many of our objections appear to rest so heavily on what *might* be, rather than what *can* be demonstrated, on what one feels rather than on what is fact. I had earlier asked Dr. Henryk Skolimowski—a philosopher at Ann Arbor—what arguments he would use to a scientist in order to try to persuade him not to pursue DNA research. He replied, "I would ask him to use his compassion." In this day and age, not many people would consider this enough. Yet it is necessary to emphasize this point. Perhaps it is similar to issues in legislative debate, such as the amount of welfare benefits, where again we cannot insist that decisions are always made on strictly rational grounds. It is the very fact of having people debate at all that is essential; differences of opinion, not matters of fact, are going to determine the outcome. There rarely is one rational answer and only one, for it isn't in the nature of most social questions. We may very learned and collect great sta-

tistics which tell us about needs, but ultimately that kind of
question is going to be determined by a feeling of *appropriateness.*

Similarly here, it comes down to a question of appropriateness again. We struggled to find some way of putting the
argument about DNA in terms of what was humanly appropriate, what was fitting. What bothered us so about the new
technology? Three things came to mind: the slow erosion of
up to this point in our history has gone to make us uniquely
human, or what we have considered to be human; the latter-
day assault on personal autonomy and integrity; and the
increasing sense that individuals are losing control over the
conduct and direction of human affairs. Time and again I
have been told—about "the technological imperative" in
general, and as it applies to this research in particular—"We
can't stop it."

Three reasons are given: first, it is in the very nature of
man to continue in this way, an inner necessity forever driving him forward to discover and know and apply. Second,
there is a self-perpetuating momentum about technology, a
momentum sustained by the beliefs and attitudes of the society in which we live; third, there is a growing existential
belief that it is impossible any more, collectively or individually, to control our affairs in any desirable way. Every time
we try to do this—so the argument goes—it ends badly, either in a way not intended, or because small groups of people
get control of the mechanisms and use them to hurt others.
The sense of being unable to exert control over one's affairs
was once believed to have been a dubious privilege of the
displaced rural poor or of unimaginative people generally,
but now it is a characteristic of a much wider range of society. The arguments of existentialism are very popular to
some, and to others, they are unattractive but the only tolerable and tenable explanations. I find them self-defeating.

Historically, it may well be that when men in primitive
societies did not understand the processes they saw and thus
felt they could not control the environment, they had to
create a religion that reflected that fact. They *had* to attribute
some mysterious or magic cause to even simple processes.
Men came to comprehend more, however, and by the late
eighteenth and early nineteenth centuries, they had every

reason to believe they had control of their affairs. It is only in the last century that we have seen the beginnings of the slow erosion of this belief. Like a suffocating miasma, the sense gradually descends: man cannot control what is going on. If you try to put the plug in one place, it inevitably comes out somewhere else, so why bother? Go about your own business, do your own thing, on your own pad, and don't put your trust in people, for they are inevitably venal. Perhaps that perception, too, is part of the new anxiety about the consequences of applied recombinant DNA technology. Man is about to be presented with a whole new set of options which *he may not really have*. Shall we do this? Shall we do the other, we are asked? And since society is still ill-equipped to make any collective decisions, others may take the decisions for us by default.

Again, if one sits back and admits that the work is probably going to go ahead, that it will roller-coaster, irrespective of what we can do, we must ask: is our reluctance no more than a small, human protest against that sheer inevitability? Since society is very good at roller-coasting, one knows that the research will go on. Nor do I have very much confidence that the molds of social direction exist yet, that either could eliminate, or at least sharply reduce, some of the possible unattractive aspects of applications of the work. We may have to create some new molds. We are excellent at supplying an infrastructure for individual scientific research. We can supply money, the laboratories, the institutional settings of the universities or institutes. But when it comes to the question of deciding *as a society* what ought to be the use or the direction that research should go, we are still inadequate. The question is, can we quickly get a whole lot better in devising a new infrastructure to enable us to handle the new science and its applications?

Yet ultimately who should make these vital decisions? Where should the power reside? A knowledgeable elite who would get data and then alone issue directives as to how everything ought to be done would be very dangerous and dispiriting. Shaw Livermore reminded me how relevant here is the story of T. H. Green, one of the Oxford idealists, who worked very hard for the second Reform Bill of 1867. When he was asked afterwards whether he believed that by

increasing the number of electors as the bill did, life in
Britain would be a lot better because people would pass good
legislation and make everything considerably more human-
itarian, T. H. Green said, "No. Those of us who worked for
the Reform Bill never believed it would result in good legislation. We believed the vote was good in itself." Green meant
that for a man to have even the first glimpse of the capacity
to control his destiny and to participate was enough. Similarly we have to strive for more than just an elite—and the
term is used in no pejorative sense whatsoever—team of
decision-makers, more than a little group of scientists or a
committee from the NIH. Somehow the ordinary person
must have the capacity to participate in the decisions, and
though it may well be that it will be impossible for everyone
to participate, at least one could hope for a level of life where
it would be a plausible possibility for the great mass of people.

The counterargument is easy to understand too, however.
Some issues are enormously complicated, requiring much
information to make an informed judgment. Indeed, the very
scientists themselves agree that even for them an informed
judgment about this research is difficult, so how could an
outsider be expected to make one? The trouble with this
attitude is that things begin to move inexorably toward a
point where such an argument becomes more and more
plausible *to each and every part of life*. So many sections of
our existence are demanding and require so much expert
information that, it can be argued, an ordinary person cannot
do any of it intelligently. Thus people get pushed out.

Yet another problem arises. The recent Karen Quinlan case
presented an appalling moral dilemma. One wonders how on
earth did society get into this situation, with such agonizing
moral choices, and the answer is clear. By concentrating on
a biomedical technology to give us the criteria for "life," and
ignoring all quality of "humanness," we *made the moral
dilemma for ourselves*. I question man's capacity to deal
with an infinite number of moral dilemmas. In doing this,
I am not questioning the capacity of man to handle situations, if he can concentrate his attention and keep everything
else in some kind of equilibrium. But we must ask: what are

we setting up for ourselves in the future with this technology? How many more moral dilemmas can we cope with?

Dr. George Steiner, a critic and writer who finds much that is exhilarating in the future prospects promised by science, also believes that they are not without menace. In a beautiful essay, "Life-Lines," he points out that the confident assumption on which Western civilization has operated since the Renaissance is now in doubt. At that time there was a belief in a natural accord between the requirements of social justice and personal worth: "Man and the truth were companions." Steiner then goes on to say:

> It is as if the biochemical and biogenetic facts and potentialities we are now beginning to elucidate were waiting in ambush for Man. It may prove to be that the dilemmas and possibilities of action they pose are outside morality and beyond the ordering grasp of human intellect. We seem to be standing in Bluebeard's Castle. For the first time, the forward-vaulting intelligence of our species, which is so intricate, yet so vulnerable a piece of systematic evolution, finds itself in front of doors it might be best to leave unopened, on pain of life.

We have had one such moral dilemma before and have managed to cope only by taking refuge in euphemisms and by using the most inhumane arguments in an attempt to rationalize our most inhumane actions. If we recall the texture of the discussions, the history of the atom bomb provides some cautionary warnings. Twenty years ago, many people thought seriously that the only way to wake up the world to the severity of the problem, the only way to get the proper international mechanisms for control, was actually to drop a bomb again. Shaw Livermore remembers lightly asking a colleague who believed this, "Where do you think would be a good place?" and being given the serious reply: "Oh, on a city about the size of *Seattle*. That would be about the right magnitude to persuade everybody in the United States that this was a serious business." Dr. Livermore said, "My God, are you really saying that the only way you can see hope in this situation is if a place like Seattle is bombed?" He was! Indeed, Dr. Vannevar Bush, that most respected of

American scientists, used exactly the same argument. In his autobiography, *Pieces of the Action*, he demonstrated a terrible hubris, a pretension to omniscience, and now, I think, a personal evasion of the real moral issue. In recalling the decision to drop the bomb, he said he would justify it because "by that time I knew that civilization faced an utterly new era and I felt that it might as well face it squarely. . . . If for no other reason I would justify the use of the bomb at Hiroshima and Nagasaki because it was the only way in which the dilemma could be presented with adequate impact on World Consciousness."

It takes a man of almost godlike self-confidence to be able to make a statement with that weight of responsibility attached. Yet it is part, perhaps even a consequence, of that stealthy dehumanization to which an unquestioning application of technology leads us. The argument takes this inexorable form: we have this new technology; it is in the nature of man to develop, advance and apply it; it has an unquestioned power, and mankind had better be aware of its extent; the only way we can demonstrate the power is to use it; therefore we must use it, even if people pay the price.

We have become so indifferent; we anesthetize our proper feelings with words. Another group at Ann Arbor is looking at world problems such as famine, overpopulation and other environmental catastrophes. Dr. Livermore showed me how, as we contemplate these things, our language becomes more antiseptic. We no longer speak of the starvation and dying of millions of people, but of "die-offs"—a term often used in animal ecology and experimention. What proportion will just "die off?" It is the Vietnam body count all over again. This comes about not only because of self-anesthetization, but through the consequences of a direct cost-benefit approach welded onto our technology, applied now not only to the Vietnam War, but to almost any administrative problem. Food gets turned into an administrative problem; cancer gets turned into an administrative problem; finally people get turned into an administrative problem. When, with some passion, I said this, Dr. Livermore replied, "Sure, and do you know why this happens? It is because it is the only successful social strategy we have ever devised for making decisions."

One soon realizes how many people, not only on university committees and in science, but in life generally, always cast questions involving decisions in the form of cost-benefit. In fact, often before we feel able to make any collective statements about a course of action, the issue *has* to be put into that form. When, as in the case of this research, the degree of hazards are unknown and it is difficult for even the experts to make an assessment, one of two things usually happens. Either people say, "Well, there will be a certain proportion of accidents, but these will be acceptable compared with the rate in industrial situations." Or they will say, "Since any one of a million things might happen, and we cannot take measurements in anticipation of any event (or organism) whose characteristics we simply cannot calculate, the question can never be gotten into a reasonable cost-benefit form." Consequently, "since we cannot make a decision ... we can go ahead."

Now, arguing that this is the *only* possible way to make a collective decision effectively legitimizes it. What we are really acknowledging is our failure to come to a decision *at all*—and some of us cannot acknowledge that failure. A cost-benefit analysis is a way of making complex issues seem very simple and rational, but it is also an absolute way of making decisions. It is as much a voluntary prison as any totalitarian jail—a prison for the way decisions must be taken. The Puritans decided the Old Testament would be the only guide to correct social action. Most of us are horrified at this, or at the very least regard it as quaint, but with our twentieth-century rationales we may be in as much of a trap ourselves as the Puritans bound by the Bible or the Marxists bound by *Das Kapital* or capitalist economists by John Maynard Keynes. Besides, the essence of a democratic society is that we do not require the people who make decisions to be of like principle, or take their decisions all on the same ground. On the contrary, a fundamental tenet in a democratic society is that there be no absolute knowledge about correct public policy. We do not say that there is one undisputed opinion and only some people who know about it. As we do *not* make that exceedingly critical presumption, it follows that a democratic system is the only political system that can proceed upon the assumption that there is no knowable principle about how public policy ought to be guided. If we say that a decision on recombinant DNA research *must* be made in cost-benefit terms, we exclude alternatives and in a sense betray the very democratic principles by which society works.

Yet how can one persuade one's colleagues? Most men of this world, whether they are the Vannevar Bushs or the Fred Neidhartsare men of gentleness and good will, making decisions not only for themselves but for what they genuinely believe to be the well-being of humanity. They would rip us apart if we insisted on deciding whether or not to go on with research simply on the grounds that we do not like the feel of things. To a great extent, this is due to well-tried academic habits of thought: the obligation not only to permit but actually to further the research of a colleague, except where the most enormous dangers or ill consequences are obvious. It is a tradition centuries old, regarded as a First Principle, a God-given commandment to dons, as it were: freedom of inquiry is sacred. All of us in academe would unhesitatingly make that choice if ever the question arose, even while we might acknowledge a risk coming from the research. But this attitude is itself historically rooted and historically variable. Very likely, if we tried, we could identify the particular factors in a particular time and place that made this value the one supreme one. Perhaps now, however, there are clouds on the intellectual horizon that suggest that this value should be questioned. Dr. Sinsheimer and others in different contexts have argued that the search for truth, far from being a God-given right, is a privilege bestowed by society.

But to raise the issue of freedom of inquiry, to challenge it, immediately provokes the strong reaction that one is behaving in the manner of the Inquisition. Look at Galileo and the Church, the scientists will say. Look at Darwin and the Church. Look at Lysenko; or at Scopes, the Monkey Trial and the State of Tennessee. Look at any of the twentieth-century totalitarian systems—do you want that to happen? That, however, is a well-known move, trying to put the other person in the most unattractive light, and worse, it is thoroughly ahistorical. "We are not in ancient bloody classical Greece," said Inspector Bones, and we are not. Nor are we in Renaissance Italy or Victorian England or Stalinist Russia.

The historical and scientific situations are *not* the same and we are *not* behaving like the Inquisition. We are at a new historical stage, with new, unprecedented powers in our hands, and new criteria and moralities are called for.

It is so easy to be sidetracked. Any kind of scientific finding has some effect on humanity, and analogous parallels for good and bad are found in history. There were the great controversies over Galileo and Darwin, and being so very careful not to be forced into the same unreasonable position, we *all* agree and smile and say, "We are not going to let that happen again." But even in saying this, we are trapped. Since we don't wish to permit ourselves to be in the same kind of reactionary conflict *vis-à-vis* science as in the past, we drive the decisions back to the scientists alone and concede them everything. We foreclose most of the grounds upon which mankind has traditionally made decisions, for we close them to the public at large. Somehow society *has* to believe, and scientists must come to be made to believe, that we *all* have the confidence to make a judgment about important scientific issues, on our own grounds and not just on theirs.

Moreover, if we insist we are fighting new battles, not old ones, and that the situation is not comparable to Galileo or Darwin, then it is crucial to try to articulate what *is* new. What is new is that with this technology, the scientists have the capacity to develop organisms which can re-create themselves: once they are made, we have to live with them. What else is new is that earlier revolutions in science, such as the Copernican, changed our perspective of ourselves in the world, or, as with the Darwinian, changed the knowledge of our origins, but this new revolution may give us the possibility to change ourselves, to alter our nature as we will, in some consciously directed form. It is not only, as Michael Rogers said in his Asilomar article, "Now that we can re-write the genetic code, what are we going to say?" It is that for the first time in history, we may have the possibility of rewriting man as we know him *out of the script altogether.* It is this which fills many of us with sadness or horror. We feel that we have neither the wisdom nor the knowledge to do this. Some of our feelings are even more prosaic. I, for one, feel reasonably content with man as he has evolved up to this time, for all his foolishnesses, stupidities and banalities

For each one of those, one can weigh a glory, a joy and an achievement, and at this stage we might be better engaged in trying to realize the potential of all people as they now are than in devoting our efforts to change man according to an ill-defined image.

There is one more point to consider. Perhaps a time has to come to round off the development of man's capacity to fiddle with his environment or himself. One is again open to the accusation of being anti-intellectual, and is pushed very close to positions with which one doesn't like to identify, but the question must be faced: can we successfully continue the slope of development of knowledge? Up to now, there have been few problems, for our capacity to cope has kept pace with new knowledge. By education and by the loosening of ourselves from the moorings of commitment to absolute ideologies, we have become much more flexible. We can handle new things, new ideas. We are remarkable. But as with the question of moral dilemmas, are we justified in expecting that our capacity to cope with these new ideas will be as infinite as our capacity to think them up?

So this is where we finally arrived after all our discussions. What began as pinpricks in our minds led to this conclusion: perhaps society is going "to have to blow the whistle," to taper off that curve of infinite environmental manipulation, as we shall have to do with infinite economic growth. We do not have a known way to do this because it has never been a problem before. We haven't even developed the capacity to *talk* about it.

Recombinant DNA may force the universities and society, as well as scientists, through some process of fundamental reassessment. In the twentieth century, the very success of modern biology has again raised for scientists the old nineteenth-century questions of wider allegiance, and this can also be seen as part of the issues involving universities and scholars. Carl Schorske, a distinguished social historian at Princeton University, in a paper called "Professional Ethics and Public Crisis" has pointed out how the very professionalization of the various sects of the scholarly community shifts their allegiance away from the local scene and the republic of letters to the narrower interests of the professional peer group alone. The social relevance of a scholar's work becomes important only to the extent that it is considered important by his professional group. Schorske goes on to argue that when scholars "broke from the enlightenment unities in the interest of substantive scholarly progress, [they] slipped unwittingly into both moral and civic irresponsibility."

To an unprecedented extent, so much of the present social crisis comes to rest in questions about the use and abuse of learning, because "never before has knowledge been the very stuff of power as it has become in modern America." Scientists may perhaps take comfort in the fact that they are not alone in being forced through a reassessment of the old scholarly ethos. Schorske concludes that the scholarly community can be protected in the pursuit of truth only to the extent that it recognizes a concomitant responsibility: a responsibility for the implications of its findings for society and mankind.

It is in these terms, and these terms alone, that one should see the whole issue of freedom of inquiry, whether in science or in other disciplines, in modern society or in "ancient bloody classical Greece."

It would be difficult to forget that evening, when with the first chill of winter some of these issues first began to come out. With gentle charm, Frank Rhodes provoked us to argue, and clearly personally delighted in the discussions, and in the scintillating display of intellectual pyrotechnics between Drs. Carl Cohen and Shaw Livermore. Argument and counterargument followed with a speed and brilliance that at times left me exhilarated, and at times disturbed. Growing in me were raw gut feelings—and how unreliable we are taught to believe they are—as I began to wonder whether all these issues were such that they should or could be settled by an up-to-date medieval disputation, or whether capacities of compassion, even aesthetic judgments, could be allowed to enter the academic arena along with pure reason when such decisions are being made. I asked Frank Rhodes that very question, to be met with the reply that has of course been the hallmark of the university tradition, "How else can we decide these questions?" Shaw, too, had had his own experience. At one session of Committee B, he asked the question:

"If we develop this research capacity, is it relevant for us in this room to talk about our *feelings*? Do we think it pertinent and proper to talk about decisions that take our feelings into view at all? Not one person of the ten in that room thought that they should.

By December faint explosions were beginning to erupt around the campus of the University of Michigan as various external groups questioned the propriety of guaranteeing the money for the laboratory facilities in advance of Committee B's report. The rubber-stamping by the board of regents only seemed to confirm the worst fears of those who felt that Committee B was merely cosmetic, for while B deliberated, Committee A had been carrying forward the business of obtaining funds and drawing up plans. There was an attitude of serene confidence or bland complacency about the work of Committee A which belied the important work being done by Committee B.

But slowly other groups of people on campus became aware of what was going on and what was implied. A group called Science for the People began its activities in the late 1960's in the Boston-Cambridge area and has subsequently spun off active chapters in Berkeley, San Francisco, Chicago, New York and Ann Arbor. Initially they were intensely radical: tired of what they felt was the smug indifference of the scientific establishment to the wider problems of science, sure that there was a political and social dimension to every aspect of scientific activity, and impatient beyond belief with the form of academic Jesuitical confrontation which they believed produces infinite talk but no action. Their initial tactics were thoroughly activist. In 1971, Hubert Humphrey was the target of a badly aimed tomato during a meeting of the American Association for the Advancement of Science, in Philadelphia. Over the years the tactics may have changed, but the aims remain the same: to expose the political implications of scientific activity with all the resources they can command; to undertake an active program of information, not only to scientists, but to the public at large; and to bring pressures to bear on scientists and scientific organizations where necessary.

Under the impetus of Professor Susan Wright, a member of Science for the People, and Professor Henryk Skolimowski, both of the Department of Humanities at the College of Engineering, a letter was sent to Dr. Alvin Zander, the assistant vice-president for research, asking that the decision to approve the research be delayed until the university and surrounding community had been informed of all sides by the best technical, ethical and legal opinions. Taking exception to the fact that so far only the position of those who advocated the research had been presented to the members of the university, they called on Committee B to face the ethical and moral implications of the research, over and above the technical questions of estimated risks. The "scientific" question of recombinant DNA had taken a new and unexpected turn.

I paid my second visit to the University of Michigan in December 1975. The counterattack had gathered momentum. But many people were extremely pessimistic not so much about the outcome but about the very likelihood of any wide public debate taking place at all. Rob Bier of the University Hospital Office of Health Science Relations was typical of them. He had for some time ghostwritten Dean Gronvall's column in the quarterly paper that gives news of the medical school, and had had more than a passing professional interest in recombinant DNA. He is an hereditary diabetic. Married, but without children, he and the children he may have would stand to gain much from any technology which would give them the promise of human insulin in great quantities, or ultimately of manipulating our genes to eliminate the ones for hereditary diabetes. Even so, his feelings were blunt and unequivocal, and he was not alone:

> After working with these people [the scientists] for three years, I find them still totally unconcerned about any issues other than the prosecution of their research. The future seems to be no further away than five years at the most. Yet when they tell me what they hope to do and I look down the road towards 1984, I think, if this is the future, *I want nothing of it, at all.* Of course I would love to be free of hereditary diabetes for myself and my children. But I would rather carry it, if all we do is to go on moving towards a scientific and technological totalitarianism.

He did not mean a scientific dictatorship, the myth of a
world ruled by Frankensteins. He was referring to the domi-
nation of technology, to technological substitutions that
come not only at the level of artificial hearts or hips and
other prosthetic devices, but at the level of mechanisms and
processes. A procreation in a test tube is for him not only a
mechanistic horror but the ultimate in the dehumanization
of a human. "To reproduce my kind," Rob said to me, "is
almost the only thing that, as an individual, was left for me
to do. Now they would take away even that last shred of
autonomy. Who the hell do they think they are?"

Soon the discussions began in the community press and
the campus papers, and it rapidly became clear what positions had been staked out. Professor Wright became the
spokesperson for the critics, and David Jackson the scientists'
representative. As the weeks passed and the tensions increased, voices tended to become emotional, even shrill. During January and February 1976, forums, lectures and
discussions took place with increasing frequency, adding to
the general weight of information and passion. Meantime a
team of scientists from the National Institutes of Health held
a site visit during the second week of January to discuss the
university's grant proposal for renovating the three research
laboratories, which Professor Neidhart described as "only a
small part of the permeating scope of recombinant DNA research expected on campus over the next five years."

This again raised the question: was the university, ostensi-
bly a democratic institution, busy creating a *fait accompli* in
favor of the research, in spite of the reservations, by now
very vocal. At the Faculty Senate Assembly in January,
Shaw Livermore emphasized that it might be necessary to
consider if society needed some method of stopping certain
types of basic research, at least temporarily. He was well
aware that "most of us have a very powerful cultural inheritance of strong support for academic freedom of inquiry," but
he nevertheless felt obliged to question whether perhaps
some type of social mechanism might have to be devised, a
mechanism not controlled by the genetic researchers, to permit selective intervention in basic research. He said that he
did not expect Committee B to include such considerations
in their report—which prompted Professor Skolimowski to

say, "Why not?" for who then would address themselves to the broader issues of the research, if not Committee B? Why was it washing its hands of such issues, especially when, as he understood, it was expressly appointed to examine them? By February, the really angry notes began to creep in, and it was at this point that the Senate Assembly unanimously decided to set up public forums for a thorough discussion of the issues.

The forum was preceded by the largest university news release ever printed, one which brought out all the views from the strictly scientific to the ethical, and it was again Skolimowski who focused on the moral issue. He wrote:

> In order to tamper with the nature of life in a fundamental way we have to have wisdom and moral responsibility; whereas in my opinion we have neither. . . . In the past we have followed too uncritically any, and every, opportunity that science and technology have offered us, often with regrettable consequences, and this myopic situation must be changed. We can't expect the scientists themselves to attack the problem because many are so immersed in their work that they don't even perceive it, and those who do probably believe that it will be against their own best interests to do so.

When the forum actually took place on March 3 and 4, the arguments, familiar to us by now, blared through the air. A highlight was the clash between Drs. David Baltimore and Jonathan King, both of the Massachusetts Institute of Technology, over the expected benefits and dangers. (This clash has been repeated at other times in other places.) King, a member of Science for the People, called the predicted health applications "a series of social-medicine half-truths," and argued that funding for a study of environmental problems— for example, those concerned with cancer—would be infinitely more beneficial. To which Baltimore, feeling that the likelihood of society accepting restrictions on smoking and industrial development was slight indeed, replied, "How much do we really need recombinant DNA? Fine, we can do without it. We have lived with famine, virus and cancer, and we can continue to do so!"

Though there was much heat, the forums did for the time being redirect some of the lightning from the acrimonious exchanges on campus, and brought the issues to a public far beyond that of the university community. Large parts of the conference were rebroadcast on the university's FM station. It was the first real public debate of the issue anywhere in the country—indeed, in the world.

That March, the report of Committee B finally came out. It supported the recombinant DNA research and said that the NIH guidelines provided an acceptable basis for it; but that the equipment and facilities should be reviewed by a committee that included at least one member of the community, preferably not a staff member of the university, and that occasional reviews be held thereafter; and that under no circumstances was the university to do high-risk work unless that restriction was specifically lifted by an "appropriate university decision-making body." They also provided, somewhat ambiguously, that "except where an experiment requires otherwise," all experiments should use only specially enfeebled organisms unlikely to survive outside the laboratory.

The report was not well received by everybody. In a further series of public and faculty meetings throughout April and May, many criticisms of compromise and failure to discuss the broader issues were raised, but in the end the regents voted to allow the research and the matter was settled. It is interesting to note, however, that the committee itself had not been unanimous. The vote was ten to one. That one was Shaw Livermore.

In a minority report, he said, "I have tried to look into the faces of those who might be immediately helped by this research, and also into the faces of those who might be overwhelmed by the capability of having basic forms of life altered." He reminded everyone that the new techniques might provide "a capability to alter life in a fundamental way," and if successful, scientists could dissolve the present formidable natural genetic barriers which separate species and directly intervene in the process of evolution.

I know of no more elemental capability, even including manipulation of nuclear forces. While it clearly would present opportunities for meeting present sources of human distress I believe that the limitations of our social capacities for direct-

> ing such a capability to fulfilling human purposes will more
> likely bring with it a train of awesome and possibly disas-
> trous consequences. Decisions will be made by individuals,
> groups, and perhaps whole societies, that may well have
> unintended but irreversible effects.

He emphasized that he did not act from a fear of the laboratory hazards, for he had been impressed with the sense of
responsibility shown by the scientists who had drawn public attention to them, several of whom were at Michigan.
And while he emphasized that freedom of inquiry must be
zealously protected in a free society, he nevertheless agreed
with Professor Robert Sinsheimer of the California Institute
of Technology, who said: "Rights are not found in nature. . . .
Would we wish to claim the right of individual scientists to
be free to create novel self-perpetuating organisms likely to
spread about the planet in an uncontrolled manner for better
or worse? I think not." Moreover, the world was not at such
a crisis point over the kind of problems that the technology
promised to alleviate that it was necessary to go right ahead
now. We could afford to suspend judgment, and, "if this
requires that humankind and science move toward a new
understanding at some point, then we must begin."

His decision was a total response of a total human being
and, as I have personal reason to know from our extensive
talks, demanded an enormous degree of courage and constant, agonized heart-searching. He was described to me by
one of his colleagues as "a very brave man"; he not only acted
with the strength of his full moral convictions in the face of
centuries of tradition, but on the one hand, was prepared to
face the ridicule of academics, their most savage and wound-
ing weapon, and on the other, to challenge the scientists on
their own ground—and believe that he was competent to do
so and quite justified.

In a letter to me later, summing it up, Rob Bier wrote how
thankful he was that the discussion never hardened into pro-
and anti-scientific lines, as it was tending to do elsewhere.
The scientists never referred to their putative detractors as
"opponents," but rather as "our colleagues who have questions." After the hearings, many people were optimistic that

the debate could be sustained at the level of sweet reason, and the possibility of a general Luddite revolt held at bay. Shaw Livermore and the dissenters did not carry the day, but their views may well form the starting point for new dialogues and new decisions, in the same way that some famous Supreme Court dissents have outlived the majority opinion. It was a small decision in one sense: to build or not to build new laboratories. But it also marked the beginning of a new socio-scientific awareness.

But the level of debate cannot always be sustained at one of sweet reason, nor has it been, and in the last year other public forums have generated considerable unease in the scientific community as to whether the nature of the current concern is really helpful or justified. Either way, public participation in some form or other is likely to be a permanent feature of the U.S. scientific scene for some time to come. Since the first public hearings at Ann Arbor, a number of similar debates or hearings have taken place: in California, New York, New Jersey, Connecticut—and most colorfully in the city of Cambridge, Massachusetts, where the city fathers called for a moratorium "in good faith." (Their legal powers in this matter are not certain!)

Color was provided by the personalities, particularly the mayor, Alfred E. Vellucci, a flamboyant character who has little love for the ivory-towered institutions of Harvard and MIT within the city boundaries, and who over the years has never bothered to disguise the fact. He called the hearings after reading an article in the local weekly paper *The Phoenix,* which revealed that though Harvard planned to construct appropriate facilities and do the work, it had signally failed to inform the public, let alone the city officials, about the likely hazards. This omission he understandably found intolerably arrogant, and it quickly rebounded on the school. His sense of theater never deserted him, whether it came out in his questions ("I want to know about these things that may come crawling out of the laboratories into the sewers"), in his comments about "monsters" and "Frankensteins," or in his actions—such as having a local choir give a rendition of Woody Guthrie's "This Land Is Your Land" before the hearings began!

Color was also provided by some interesting new testimony, perhaps highlighted by the revelation that the Harvard laboratories were infested with cockroaches and nasty red ants that pick up background radioactivity and that so far have resisted all the efforts of science and technology to exterminate them. Where science has failed so dismally in keeping these contained, how, it was asked, could we have any faith in its capacity to keep recombinant DNA organisms contained, P3 facilities notwithstanding?

To the by-now-familiar figures who regularly give evidence, with their by-now-familiar arguments, others were added, notably George Wald, Nobel Laureate, and his wife Ruth Hubbard, both of whom have come out strongly for restricting the research to a few special places like Fort Detrick that are best equipped to handle such problems. Dr. Wald considered the issue the largest ethical problem science has ever had to face. "Our morality up to now has been to go ahead without restriction, to learn all that we can about nature. But restructuring nature was not part of the bargain. "We are not," he argues, "simply free to mess around with billions of years of evolution." The whole issue should be more widely debated on a national and international level, he felt, before anything irrevocable was set in motion.

Once again, recombinant DNA chalked up another first for science: for the first time in history, scientists were called on to defend their work in the full glare of spotlights—the hearings were televised—and with the fullest local and even national coverage. At the mayor's summer fair in July, both the scientists and the environmentalists set up booths to explain the benefits and dangers of the work to interested passers-by. Then a second three-month moratorium was called, during which a citizen's committee appointed by the city council was asked to examine the question and make recommendations. Its final report was, in my opinion, superb. It grasped the nettle of public participation firmly and stated the principle unequivocally: it *is* possible for lay people to educate themselves properly on scientific issues, to grasp the technical complexities as well as the ethical ones, and then to make fair and rational judgments. Their report, which was finally adopted by the city council in February 1977, recommended that Harvard be permitted to proceed with the construction of a P3 facility, that all such work in the city be registered, and that a laboratory experimentation review board keep a watching brief on the work at the local universities. (Appendix II is an extract from the committee's report.)

Such public exposure and input is presumably everything that Senator Kennedy, for one, could hope for, and many people do indeed hope that the Cambridge hearings will be a model for other such hearings around the country. Mayor Vellucci himself introduced a resolution at the United States Conference of Mayors, to the effect "that no university or institution may begin recombinant DNA experiments without first notifying the mayor of the city or town in which the experimentation will take place. Thereupon, the mayor and legislative body shall call a public hearing so that the public may be an active participant in the decision-making process." The resolution was deferred to a committee for consideration. At the state level, too, legislative moves are contemplated: in Ohio, Wisconsin, Indiana, New York and California. Governor Jerry Brown, for instance, is to introduce legislation in California to require that all universities and industries conform to the NIH guidelines, as well as any work that is done under patent procedures, whether private or public. Once again all work within the state will have to be registered and open to inspection. The Hastings-on-Hudson Institute will have an indirect but significant influence in the person of Dr. Mark Lappé, who is director of a new office (Health, Law and Values) in the Department of Health of the State of California, for Dr. Lappé was one of their earliest and most effective staff members in the area of genetics.

Good in principle as these moves may be, however, and though more and more scientists have come to accept the inevitable slowing of the pace of the work, there is some entirely justified uneasiness on the part of the scientific community. At a simple level they do not want a whole series of local options to be available, for it would be inequitable and chaotic if experiments were banned, say, on one side of the Charles River but permitted on the other side. As it is, this is the situation on both sides of the Atlantic. A number of distinguished American scientists have received invitations from Switzerland to go there and do experiments which they cannot easily do in the States. These days the Swiss seem to be regarding any escaped microbe and any shackled scientist as political refugees.

There is also concern that a mandate to look into all laboratories and at all research could end up as a scientific witch hunt. There is equal concern that the whole flavor of the discussion, which privately has at times been venomous, could intensify the fires of the anti-rational, anti-scientific movements and bring down the whole temple of science about our ears. Concerns such as these surfaced especially after the Cambridge debates, where defenders and opponents alike were brilliantly articulate, highly motivated and politically experienced.

So now we are walking a tightrope, as we struggle to preserve those activities and values that, in the past, have been a pride and an ornament to Western society. Yet we must also seek to control our knowledge and insist that whatever is done occur both for the well-being of society and with its widest consent. Anything less cannot be contemplated.

Conclusion:
Walking
the Tightrope

I pondered how men fight and lose the battle, and the thing they fought for comes about in spite of their defeat, and when it comes turns out not to have been what they meant, and other men have to fight for what they meant under another name.

—William Morris, before listening
to a sermon from John Ball,
in his vision of the lost past

It was difficult when writing this book to avoid sensations of schizophrenia, and there was a time when I felt that I went around bearing the imprint of the last person who sat upon me. There are so many points of view, and almost all of them are deeply held and convincingly justified. I came to have a real sympathy with the Cambridge city council member who, after hearing conflicting testimony from several very distinguished scientists, burst out, "What the hell am I supposed to believe?" Nothing is known with sufficient certainty yet to enable us to come to reasonable conclusions. We remain forced to consider a battery of quite *distinct* questions and alternatives. The safety of the procedures is an issue a world away from the ethics of human genetic engineering, yet together they are caught up in a wholesale reassessment of the place of science and technology in our lives.

Of course there may be a joker in the pack. Dr. James Watson believes that we may all be playing a role in a black comedy, creating a scare that does not really exist and working ourselves up into a macabre and dangerous tizzy. Perhaps there *is* no problem; perhaps in procedure and application, recombinant DNA will be perfectly safe and admirable and as innocent as the driven snow. Perhaps this episode will be the biggest nonevent in twentieth-century science. But when I suggested this to Dr. Charles Weissmann, I provoked another of his old-fashioned looks. "Not to us—within science—it isn't," he replied. "It could never be."

We shall just have to wait and see, for this tale is by no means done and it is unlikely to end in the well-rounded fashion of fiction. Situations in real life rarely do. What we are seeing is both a sudden change in the dynamics of scientific progress and the opening bargaining moves in the process of coming up with a new social contract between this remarkable profession and society. If there can be one conclusion at this stage, it is an obvious and irrefutable one: things will never be the same again. But it is much too early to tell quite where they will go from here. Historical prediction has not been very successful, both in science and society, particularly when one is dealing with a rapidly shifting historical matrix. A delightful lecture by Professor Asa Briggs, the social historian, entitled "The Nineteenth-Century Visions of the Twentieth" reminds us of history's many "cunning passages" and "contrived corridors": there is hardly one single prediction made by our Victorian forefathers that compares with our present reality!

If indeed we are witnessing a rewriting of the social contract between science and society, we need now ask two further questions: What form should the new social contract take? What form is it realistic to expect it to take?

We have seen in this book several of the reasons why society demands that this contract be rewritten. We are no longer prepared to tolerate super-bureaucracies which act in the name of individuals on the assumption that the technical bureaucrats know what is best. How long the present mood will last, of course, is another question. American society and American academic life have long been characterized by these enormous swings in concerns, interests and fashions. Great new theories are regularly produced which at the time embrace new attitudes and beliefs in a tremendous comprehensive sweep; questions and their solutions become swept up into systems of thought which rapidly acquire the mystique of eternal verities. But these trends never last long, and it may be the same with the antiscience movements. If science is merely the last in a long line of public activities to endure this scrutiny and reaction, the heat may eventually be taken off, especially if it can chalk up one or two great successes, such as an effective understanding of cancer, with concomitant therapeutic application. Meantime, however, the new social contract remains to be negotiated.

If a new social contract is to be written, one would ideally like it to specify several things. As philosopher Max Black said in 1975, when addressing the American Association for the Advancement of Science, there can be "no return [for scientists] to that state of moral innocence which characterized earlier periods." This state was "a vestigial survival of an earlier and much safer era." It was safer not only because science was a more protective self-contained system, since its application on the whole seemed to work for the good; it was safer because the society in which the profession operated was quieter and more secure. By insisting now, in the turbulent years of this century, that scientists can no longer claim moral innocence, we are asking them to exercise their imaginations in new areas, and we have to insist that both scientists and their institutions take more of a lead in this: to use their moral imaginations to try and calculate what potential dislocations in society or physical and phychological risks to individuals would result from doing certain experiments or applying their results. In effect, scientists are being asked to think through in advance not only the consequences of certain experiments, but indeed their very choice. To give one small example: it would be enormously useful both from the point of view of the profession and of society if some scientists settled down for the three or four years it would take and studied the ecology and epidemiology of the bacterium *E. coli.* This knowledge would be vital both for an assessment of the actual risks we may be running and for an understanding of just what is possible in genetic engineering and of the relevance to man of certain experiments in mice. Not many scientists are willing to take the time to do this, however. That field is not "hot"; it is unlikely to yield immediate and striking results, because ecological studies are inevitably very complicated. A scientist looking for a quick result and a quick return to help his career will seek a fashionable field with questions that yield quick answers.

For many reasons, it may be unrealistic to expect science and scientists to change their ways and their habits of thought overnight. For a start, they don't want to. In the last eighteen months, I have had many biologists remind me of J. Robert Oppenheimer's phrase spoken after the Manhattan Project, "Physicists have known sin"; biologists quote this as an historical precedent to justify their present collective attitude toward recombinant DNA and its applications. They argue that if it is now their turn to "know sin," this is a price they will happily pay for going on with their own work.

Moreover, we are asking scientists to take time out from what it is they enjoy doing most, so as to take effective political actions to educate the public on their aims and the outcome of their research. We are insisting that they show a new degree of moral responsibility and make moral choices. At the same time, we are warning them that we will be monitoring their activities, calling them to account in a wide variety of ways. We should be aware of just what this implies for the process of scientific discovery, and ask whether this matters. One reason why the Asilomar Conference turned so quickly into a technical meeting was not only because that is the kind of activity which scientists can handle easily, but because forcing them to jump through a series of political, moral and social hoops is utterly destructive of the creative process of doing science. It is dislocating; it is frustrating; it seems to have few relevant end-points—if any end-point at all. A real creative scientist needs to put on the blinkers and be undistracted by *anything*. Only those who are capable of total immersion in a problem and who can look at one situation in one hundred and fifty different ways can hope to come up with an imaginative, let alone a correct, solution. They have to close the door and hole up.

If society wants good science and the results of good science—and it apparently still does—it must face this creative process. If we had asked Pablo Picasso, Robert Frost, Aaron Copland or Benjamin Britten to do their art with an eye to the wider considerations and effects, to take time out to educate us, and to take effective political action, there would have been screams of outraged frustration. The core of creative activity is the same for art and science. This is why one would hope that the *institutions,* the professional bodies, would take over some of the public and professional roles that we are asking individual scientists to assume.

What can we reasonably ask of the institutions of science? For a start, they could actively encourage all those practical steps which, by producing concrete evidence, would alleviate some of the hypothetical fears about recombinant DNA technology. These include doing experiments of the sort suggested by Dr. Norton Zinder, experiments that would establish just how, and when and with what effect DNA is—or is not—transferred and expressed across different classes of organisms. Senator Jacob Javits spoke of the possibility of a "biological holocaust." Neither he nor we can yet know whether this is sheer alarmist exaggeration or within the realms of possibility. But we could start to find out. Second, a crash program of research could be mounted to work up another bacterium to replace *E. coli.* If so many of the dangers appear to rest on worries about the spread of this bacterium in man, then many people—including some scientists—find its continued use indefensible, except on the grounds of scientific self-interest. A large industrial organization could easily mount such a crash program. *Not* to do so is, as Dr. Margaret Mead told the president of Hoffmann–La Roche, Inc., "pure self-indulgence." In addition, a start could be made, as Dr. Paul Berg recently suggested, to use identifying markers other than antibiotic resistance for these bacteria. Dr. Sydney Brenner has described the world as a "dilute solution of tetracycline," so widespread is its use, and strains of bacteria resistant to it can be expected to multiply fast.

Moving to the wider, less technical issues, the institutions could sponsor, encourage and even provoke wide-ranging discussions about human genetic engineering and the wishes of society. For example, the joint forum of the National Science Foundation and the National Endowment for the Humanities would be an excellent stage for this. Internally, the institutions could encourage the creation of courses in the scientific curriculum designed not to do with the philosophical relationships of scientific theories and their technical applications, but with their ethical and moral impact. Dr. Jerome Ravetz has written that the average scientist entering the profession is as well equipped to cope with the ethical and moral issues they may be called upon to face as is an adolescent Irish girl arriving at Euston Station. People taking up the profession of science should be aware of the social environment in which they will be practicing science, of the forces and the pressures upon them, as well as the forces and pressures that their work itself will generate. They should know the ethical and moral dilemmas in science, and they should be aware, too, that they have both a power and a *choice:* in deciding to go into one area of research rather than another, and once there, to go for one kind of investigation over another, they are making decisions *on behalf of society* as to what shall be investigated, and in what form.

At the Airlie House Conference, called by Senators Kennedy and Javits in 1975, a number of scientists, politicians and philosophers spent two days discussing how to give the public some say in the act of scientific choice. Dr. David Baltimore suggested that there be some mechanism whereby the members of the public could draw up a list of problems on which they thought scientists should be working, for scientists to consider. He then went on to say that probably most of the problems would be already in progress ... but I wonder? The ecology of *E. coli* is one case in point, and how many scientists are working on the problems of malaria, schistosomiasis or leprosy for instance? As Dr. Sydney Brenner and Dr. Barry Bloom have separately reminded us, compared to the medical problems that recombinant DNA will help, the world-wide problems of human parasitic disease are overwhelming. Nevertheless, Baltimore's suggestion would make a good starting point.

Public understanding of science is another extremely crucial area for the scientific profession. It has great relevance for the question of what society expects of science and needs it for, but little is being done to foster that understanding, and most of what is being done is pitiful. A scientist who chooses to take time out to explain his work is penalized in his profession; so, too, is an academic who instead of writing learned articles for learned journals takes time out to write popular books. Equally, I know many newspaper reporters who would be delighted to be given the opportunity to be free from the acute editorial pressures of newsworthy headlines—headlines which can lead to the distortion of scientific information. They would welcome the chance to write about science at leisure, but again, few opportunities exist.

Lastly, one could reasonably expect the institutions to take an infinitely more positive role as protectors, not only of science itself, but of the public interest so far as the application of science is concerned. This issue was highlighted by a recent report on "Scientific Freedom and Responsibility," written by a committee chaired by Dr. John Edsall of Harvard University and issued under the auspices of the American Association for the Advancement of Science. Three recent episodes in their report are particularly instructive. In two—the cases of two doctors at the Atomic Energy Laboratories in Livermore, California, and of three engineers working for the Bay Area Rapid Transit system (BART) in San Francisco—the scientists pointed out the hazards of certain procedures to their employers. Because they disagreed with the public statements and were prepared to say so, they were harassed, and in the case of the BART engineers, dismissed. In a third episode, scientists working for the Manufacturing Chemists Association—the association that would not testify before Kennedy's panel—suppressed information about the dangers of vinyl chloride, even though they were well aware of them. They did not even notify the National Institute of Occupational Safety and Health (NIOSH), let alone warn the public or the workers exposed. In January 1973, NIOSH requested information on the possible hazards associated with twenty-three chemical substances, including vinyl chloride. On March 7, the Manufacturing Chemists Association responded, but merely recommended that a precautionary label should be attached to the products. They made no mention of the toxic effect on animals or people. In the words of John Edsall, "The association appears to have deliberately deceived NIOSH regarding the true facts." The MCA claimed that the withholding of information was due to an agreement with European manufacturers to keep the data confidential until the wording of a release could be agreed upon. However, as Edsall points out, because of the suppression of this data, tens of thousands of workers were exposed without warning for perhaps two years to toxic concentrations of vinyl chloride.

So we have two problems: on the one hand, scientists being penalized for acting in a socially responsible manner, and on the other, industrial scientists and their institutions acting in a downright amoral manner. An effective professional body could have avoided both situations, and it would seem not unreasonable to ask that professional societies now take an active role. They should defend their members who attempt to act in the public interest; they should collectively and individually bind themselves to permit no interests, economic or political, to stand in the way of the scientific truth as they see it. As Dr. Edsall's report points out, most societies have in the past remained aloof from conflicts of this sort. They have taken the attitude of the nineteenth-century British association: the purity of their devotion to the advancement of science would be contaminated if scientists entered the public arena, for political expediency might then tempt a scientist to distort his results. This is precisely what has happened already, however, as a result of the profession's political detachment. Such attitudes are no longer appropriate or permissible.

If the professional institutions could move to anticipate, or at least come to terms with, the public's very genuine fears and rightful concerns, they would not only be doing an enormous social service; they would be acting in their own best interests as well. This kind of professional stance would save individual scientists from the kind of public harassment that Dr. Joshua Lederberg so obviously feared. When a scientist entering the profession knows that he binds himself to conform not only to a methodological ethic and the truthful reporting of results but also to a moral conduct of his work

and its application, then and only then could one say that the
scientific profession had achieved a full professional status,
comparable to their colleagues in the law, for example. The
institutions of science would then act as guardians, not only
for their own and each other's competence, but for the public
interest.

The onus is not only on the scientist and his profession,
however, though some people behave as if it is. As we have
seen, science emerged out of society amidst great public indifference. As time went on, however, society realized that
it wanted science and used it over and over for its own ends.
It will probably continue to do so. Once society's own myths
and illusions about the activity have been stripped away, it
might even, indeed, come to love it again one day! J. D. Bernal
once said, "A strategic mistake may be as bad as a factual
error." Society as well as the profession has made bad strategic errors, and has rarely bothered to examine the nature of
the scientific activity in any depth, nor its own relationship
to the profession. Senator Kennedy has quite rightly insisted
that the scienfific profession be more acutely aware of the
public's needs, but what are those needs, and have we ever
sat down to work them out? What do we really want science
for? For technology, civilization or both? The mechanisms
that would enable the public to come to some kind of collective decision about these questions have never been set up
or even actively thought about. Again, the impetus could
come from the scientific institutions, with, one hopes, congressional or state support. Perhaps the same mechanisms
for public participation that have operated in the Concorde
debate could be resurrected, providing scientists with ways
and means of sensing public understanding, but concentrating now on the *applications* of science rather than on the
discoveries. For instance, Paul Berg suggested to me that
there should be a public conference designed to find mechanisms other than those that involve human genetic engineering for dealing with those medical problems that at this
stage of our knowledge seem to require human genetic engineering. He is confident that science can steer a course
through such technological minefields. These forums will
have to be much more closely interwoven with the political
and legislative procedures. All this may slow up scientific

progress somewhat, and again we must ask ourselves: do we mind, and does it matter?

What is really at issue is the enhancing of collective ethical sensibilities in society, both in professions and in individuals. We are really talking of that most difficult problem of all: creating new moral climates in which social sanctions alone may be sufficient to impede the wholesale inhuman application of a technology by anyone, scientists or manufacturers, psychosurgeons or advertising men. We would do well to cease to frighten ourselves with unreal visions of wicked Frankensteins creating new world orders. The real dangers and fears are minor in quantity, but are more likely and just as outrageous. The objective application of a technology to even one human being that does not bring benefit to that human being is the real sin. The scientist or politician —indeed, anyone who acting in his own interests does something that affronts the integrity and dignity of another—is the person we have to guard against, and against whom our moral outrage should be expressed. But we must beware of hypocrisy here. For we have no right to insist that one profession conform to moral standards that we as a society do not follow.

At the very core of this debate is the issue of power and public control of this power. If we have not yet developed the tools for this control, it is because we have only just become aware of the problems. Groucho Marx has said, "These days, it takes an optimist to be an optimist," but one of the comforting conclusions I draw from the whole of the story of recombinant DNA research is how remarkably fast a democratic society can respond to an obvious problem and can begin to develop procedures for dealing with it. We do have grounds for optimism. Maybe we move too slowly for those who are frustrated with Jesuitical confrontations, and our action is too thinly spread for those who, constrained within an ideological framework, would impose a total decision on behalf of others. But the system of checks and balances built into the British and American democratic systems is something in which we can take justifiable pride and which, so far as the new relationship between science and society is concerned, is beginning to work tolerably well.

It is in everyone's interests that it should, for science and democracy are really very similar institutions whose very conditions for existence are the same. Both are equally fragile. They require us to walk a tightrope between an excessive authoritarianism that stifles imagination, and an unbridled license that can result in an intellectual or political anarchy. The danger will only come, as Sir Peter Ramsbotham reminded me when we were discussing the nature of social change in Britain, if we lose our tempers, or if we fail to appreciate the quite justifiable concerns and interests of others. As I researched this story, it was sad to see how many people still thoroughly mistrusted their opponents' motives, and how angry and vindictive some people are. Even to use the term "opponent," however, is to muddy the waters unnecessarily. To regard this process—hammering out a new social contract—as a battle, with victory to be won or lost by either side, would be to do a great disservice both to the issue and to the way we must deal with it. Some people undoubtedly do see it in terms of outright conflict, but they would do well to remember William Morris's reflection at the beginning of this chapter: battles are almost always lost, yet the end-result comes about finally—though not quite in the form the proponents expected. Now, more than ever, both inside science and outside, we are in this situation together. Those who have been vehemently antiscientific can take great comfort from the number of young scientists who said to me as I worked on this book, "If society as a whole decided that recombinant DNA research should not go on, I would kick and scream, fight and swear, go off to Timbuktu or change my problem. But in the last analysis I agree that society has a right to make this choice, even if it is an erroneous one." So I feel it is right to give scientists the last word here. Dr. Peter Carlson once said to me, "We are all human beings. We are in this world together and we have to work it out together. When we impose our constraints on certain areas of society, be it industry or government, or science, the rule of law of the people has to come first. *They* have the first priority. People, not ideas nor things. What we have been lately doing in science and technology has been removing the spirit of humanity—taking away some of the last measures of humanism. This we have to reverse. Even the scientific quest became more of a social quest, something that you could make into a technology, something that worked, something that made a profit. We must change all this too. Go back to it as a personal quest and revive a lot of the spirit of the earlier ages, both with regard to the activity of science and to our love and understanding of the men within it."

Afterword

One's belief that one is sincere is not so dangerous as one's conviction that one is right. We all feel we are right; but we felt the same way twenty years ago and today we know we weren't always right.

—Igor Stravinsky

In March 1977, just two years after the Asilomar Conference, the National Academy of Sciences of America held an academy forum on research with recombinant DNA. From time to time the academy holds such open meetings, designed "to aid in the resolution of dilemmas arising from the discoveries of science." This particular forum was an important meeting and brought together many people from the international scene, not only scientists but also politicians, lawyers and philosophers, as well as members of the press and television. It provided an ideal occasion for a summing up of the current situation, as well as an opportunity for an exchange of public and professional views. Personally I was given a delightful opportunity to meet again many friends I had made in the course of writing this book, and to get up-to-date with their views. Yet at the same time I was reminded of the final chorus of a great theatrical spectacular, the moment when all the cast assem-

bles on the stage for the reprise and the nostalgias of fare-
well. For this forum will, I believe, turn out to represent the
moment in this story when attention finally swings away
from the issue of whether or not to do the research—which
is now decided*—and of safety to questions of the public
input into science, and the applications of the technologies,
especially as they relate to human genetic engineering.

There was an abundance of reprise, for practically all the
characters in the drama reappeared, and with minor variations, the main themes were recapitulated once more. But if
there was any nostalgia it was not for the present drama, but
for the golden days long since past, when the practice of
science unfolded quietly, away from public dissection, the
time when man and truth were allies.

Like Asilomar before it, this conference had its own particular flavors. One sensed the incipient trauma in many established scientists as they faced—so it seemed to them—a
dangerous situation of science under strong attack. They are
uncertain of how such a situation came to pass, uncertain
how to react, and in turn aggressive and dismissive of opponents. They feel it bitterly unfair that their motives and
actions are so misunderstood, if not downright misrepre-
sented, and some of them clearly interpret the whole episode
as an awful warning of what happens when scientific disagreements are debated out in the open. People will believe
a simple lie rather than a complicated truth was one sad
comment. Their instinct now is to "hole up." In future, if
there are disagreements within the profession, these should
be kept under wraps until debate or evidence finally resolves
them. Yet others—and it is difficult to assess whether they
are in the majority, but I think it unlikely—wisely feel that
what has happened is so much inevitable history, that there
can never be a reversion to the salad days, so it is in everyone's interest to accept the current situation and work from
within it. For in any case the distinction between public and
private science is no longer clear.

The sheer fatigue of many of the participants, especially
those of the initial organizing committee, was strikingly

_____

*It may not be . . . if the Friends of the Earth are successful in their court
action to ban federal funds for this research.

evident. And whatever one's views about the pros and cons of the debate and personal desires for any one particular outcome, it is impossible not to feel deeply sympathetic towards them. They are people of integrity and for two years now have, without respite, been called upon to repeat their arguments in a variety of places, to a variety of audiences, on television and in print, facing a wide spectrum of antagonism, support, skepticism and confidence. For them this show has had a much too successful run and by now they are just plain tired: tired of being misunderstood, tired of being attacked or vilified, tired of having to explain the same thing over and over. But this is what politics and much of everyday life is like. It just so happens that scientists have had little experience of the political arena. But from the moment the issue of public safety for recombinant DNA was first raised, the research became a political matter. So one may sympathize with them while not being surprised at what they now face. However much they long to get back to a quiet problem in a quiet laboratory, it is clear that the debate has not yet run its course.

The work will now definitely go on. This much is clear. But what will alter our views about its dangers or its benefits? Only hard empirical evidence, and it is plain that some of this is beginning to emerge. For instance, the laboratory strain of *E. coli*, EK12, has now been tested in a number of human male volunteers and so far has not managed to survive at all in their intestines. Though the construction of the experiment angered one woman scientist at the forum, who felt that it should be extended to the other half of the human population, it is unlikely that the outcome would be any different. On the same issue, Dr. Roy Curtiss of the University of Alabama spoke about his new version of the weakened strains, numbered 1776—again after the Bicentennial. He has built a constellation of crippling mutations into this version of *E. coli*, which obviously hampers it most effectively for life outside the Petri dish. For, amongst other defects these mutations prevent the bacterium from passing on its genes to others, make them lethally sensitive to ultraviolet light. As a result, *E. coli* x 1776 can live neither in the human gut nor in human blood serum. Reminding us that we are dealing with a situation of low probability but of high risk, he called for experiments to determine the fate of the DNA of the plasmid. Does it get across the intestinal wall, or does it not? For the real problem was not one of infectivity or of pathogenicity, but what actually happens to the information that the plasmid is carrying. The Catch-22 is that there is only one way to find out: do an experiment. Though at the time he emphasized the need to discontinue the use of antibiotics as markers for the plasmids and bacteria, his main recommendations were not technical so much as human, for the potential for human error provided the highest risk. There should he said, be nonscientists on the safety committees, for like housewives advising architects who design kitchens, they would be able to spot things that scientists, too familiar with their surroundings, would miss. Like good housekeeping, laboratory procedures should emphasize scrupulously clean methods; only then could rodents and insect contaminations be avoided. The red ants of Harvard have made their impact. The Cambridge City Council has insisted that the university guarantee that its P3 facility is free of insects. But because it is housed in an old building, that may be the one condition the university cannot meet. In the same way that a rare species of snapdragon, living in the valley of the St. John River, Maine, stopped the Dickey Lincoln Dam Project, the ecological impact of the red ants of Harvard may prove to be more effective than all the protestations of George Wald and Ruth Hubbard.

On the plant biology side, Dr. Ray Valentine of the University of California at Davis provided good empirical information with a light touch. He revealed that at Asilomar, Paul Berg had asked him: "Will the nitrogen fixation (NIF) genes escape?"—i.e., can they extend their range widely and upset the usual ecological balance? Dr. Valentine has spent a great deal of time in the last two years trying to answer Berg's question. He concludes that they not only *would* not escape but *could* not escape, for they require much energy to function at all. This energy has to be pulled in from some source, and this effectively limits the conditions where they can live. Indeed, this will probably be the main limitation on their potential as a source of fertilizer. Though people have tried, they have not managed to get even one nitrogen-fixing gene attached to a cereal crop. But there would be other ways around the fertilizer problem, and these might not involve recombinant DNA at all. He showed us a photograph of rice paddies with stretches of a vivid green water between the rice plants. The color was due to a blue-green alga, *Azolea,* which thrives there happily and gives fertilizer to the plant indirectly via the water. It was a good source of protein as well as of fertilizer, and with the air of a magician producing a rabbit from a hat he proudly waved a flask full of the plant and presented it to Dr. Berg. It has, Dr. Valentine told us, a nice nutty flavor, and with a Thousand Island dressing slides down very well.

We were brought up to date with the possibilities of genetic engineering. The news from Cambridge, England, had just arrived: Dr. Fred Sanger's group has specified a virus, genetically and biochemically, in full, for the first time. "We've now got the gene," someone told me. This was greatly exciting and brought nearer the day when human genetic engineering would be possible. Reviewing this subject, David Baltimore said that though it would be many years before the human germ line could be altered, within five years or so, he could forsee human genetic engineering in the following form: for example, to help people who have a hemoglobin deficiency of some sort. The bone marrow might be removed from a child with this deficiency. The correct human gene could be replaced by attaching it to a plasmid, which would then infect the bone marrow cells. These could be replaced in the patient with their function restored, and since they are the person's *own* cells, the problem of rejection, presently the most serious problem of bone marrow transplants, is completely avoided.

Politically, the situation evolves towards the long expected end-point. On April 6, Kennedy held legislative hearings once again. Indeed, a few days before, the senator had introduced a bill into the Senate containing elements of others earlier proposed by Dale Bumpers and Maurice Metzenbaum. At the hearings the Secretary of Health, Education and Welfare, Joseph Califano, said that the administration proposed introducing legislation to cover all recombinant DNA in the United States, whether private or public. This would have to be licensed and there will be civil penalties brought against those who evade or transgress. The one obviously contentious—and presently unresolved—issue is the one of public scrutiny. Once again industry is fighting this, insisting that it is not in commercial interests that they should be required to reveal everything they are doing.

At times during the forum one felt that the whole story was neatly rounding off. Actually it is not. At least two important issues remain and will be with us for some time. One is the problems raised by human genetic engineering. Everyone could sense and most all agreed that this issue, while intellectually quite distinct from that of the safety of the research, is nevertheless bound up with the work. It is the one issue which arouses most debate, the most emotion and the most anxiety. Two distinguished academicians urged their colleagues to recognize that over this question, emotive gut reaction was not only understandable but possibly perfectly justifiable. To ignore, dismiss or evade it would be inhuman and morally quite wrong. The next debate will come here, and for the first time we may at last have a full examination of an issue in advance of a technological application, and work out procedures and institutions needed for dealing with the new scientific powers we have. For a start, it is quite likely that there will be an open commission to study these questions. I myself foresee not just a positive advance here but the likelihood that by these means, many of my own unhappy fears may be set at rest.

But the social contract between the scientific profession and society, the point from which I began this work, and the second outstanding issue, is quite another matter. After this forum I am not too optimistic here, for the scars of this episode have been etched deep. One of the most poignant moments for me came when questions central to this issue were being discussed. David Baltimore spoke with regret of two things; one, his own counterargument to Jonathan King at Ann Arbor, when in fiercely defending this work he had said, "Sure we can go on living with cancer." It was not, he said, a remark that he is happy to remember. But he spoke nostalgically of his earlier days in science—a time when people equated science with the search for truth, and both held science in regard for that reason and wanted science for that reason. If society had changed its stance on this, then something very fundamental had been altered in the intellectual world—and the price may be a high one.

Yet the question of scientific and social priorities remains. Many of the opponents of the research have based their attack on an extension of the question "Will the expected benefits outweigh the risks?" "We do not need this research," they said. Take cancer: it is clear that an effective program of prevention will do more to cut down the terrible consequences of this man-made epidemic, and much faster, than recombinant DNA research can do. This is quite true. Equally, with the food problems of the Third World or the fertilizer problems: simple social and economic remedies are already available which again will meet the problems just as effectively and again, much quicker. Moreover, there is no crisis or problem that the world faces of such critical immediateness to which recombinant technology could be applied, that we must steam full ahead. The situation is totally unlike that in World War II, when it was possible to argue that because the Germans might be working on an atomic bomb we faced a crisis to which an atomic technology had quickly to respond. At the root of these objections are deep moral concerns about the priorities of our society, and as with the social contract, recombinant DNA focused and intensified these questions. They are quite irrelevant to the questions of safety of the work, but during the debate the various issues became thoroughly merged. There is no question that both our society and the scientific community have self-centered priorities, and it is both scientifically possible and morally right that a significant proportion of our resources, both intellectual and financial, should be turned towards the pressing needs of the contemporary world. Here, at present, recombinant DNA cannot have the same impact and to present the research as a basic source of solutions to these needs is not right. Once again, I feel this is a situation where the institutions of science, as well as its members, should take an effective lead in seeing that these aspects receive their rightful scientific and economic attention. But I am not sanguine about this.

Another moving moment came with the final address by Dr. Tracy Sonneborn, a gentle, much loved scientist, distinguished for his work in genetics. It will be, I suspect, people

like him who represent the very best we have believed about
science who will help heal the wounds between the scientists themselves and the profession and society. His was an
address through which simple moral threads ran like pure
gold. Others had pleaded with people to take the heat out of
the debate, especially to reduce the personal acrimony and
venom—to keep personalities out of it. Dr. Sonneborn evoked
the only worthwhile nostalgia. He recalled the old days not
because they were times when everyone went on doing
"their thing" irrespective of others' views or feelings, and not
because of the unlimited money and power, but because
differences were expected, and when they arose, they were
dealt with by gentlemen. Contentious issues did not on the
whole evoke personal and open hostility but stimulated research to narrow the areas of disagreement, if not find the
answers. At times during the last two years I suspect he has
not recognized the profession to which he belongs.

It is difficult not to respond to a simple moral force—at
least temporarily. An example of this nature will not change
a situation overnight, no more than Jimmy Carter will suddenly restore the moral tone of America, but it serves as a
necessary and quiet reminder of very simple truths. This
episode of recombinant DNA could be the starting point for
a new social contract both within the professions and in its
external relations . . . a contract where—as Rousseau said—
each person gives up a measure of his freedom in return for
the protection of freedoms for all, so as to maintain a desirable human condition. But will this opportunity be taken? I
simply do not know.

Appendix I:
The NIH Guidelines

The guidelines define four levels of physical containment, designated, in order of increasing stringency, P1 to P4, and three levels of biological containment, EK1 to EK3, and assign experiments to them on the basis of potential risk. The following is a summary of containment levels specified for various sources of DNA.

a. Shotgun experiments using *E. coli* as the host

Non-embryonic primate tissue	P3+EK3 or P4+EK2
Embryonic primate tissue or germ line cells	P3+EK2
Other mammals	P3+EK2
Birds	P3+EK2
Cold-blooded vertebrates, non-embryonic	P2+EK2
embryonic or germ line	P2+EK1
If vertebrate produces a toxin	P3+EK2
Other cold-blooded animals and lower eukaryotes	P2+EK1
If Class 2 pathogen,* produces a toxin, or carries a pathogen	P3+EK2
Plants	P2+EK1

Prokaryotes that exchange genes with *E. coli*	
Class 1 agents (non-pathogens)	P1+EK1
Low-risk pathogens (for example, enterobacteria)	P2+EK1
Moderate-risk pathogens (for example, *S. typhi*)	P2+EK2
Higher-risk pathogens	banned

Prokaryotes that do not exchange genes with *E. coli*	
Class 1 agents	P2+EK2 or P3+EK1

*Classes for pathogenic agents as defined by the Center for Disease Control.

Class 2 agents (moderate-risk pathogens)	P3+EK2
Higher pathogens	banned

In all above cases, if DNA is at least 99% pure before cloning and contains no harmful genes, either physical or biological containment levels can be reduced one step.

b. Cloning plasmid, bacteriophage and other virus genes in *E. coli*

Animal viruses	P4+EK2 or P3+EK3
If clones free from harmful regions	P3+EK2
Plant viruses	P3+EK1 or P2+EK2
99% pure organelle DNA, Primates	P3+EK1 or P2+EK2
other eukaryotes	P2+EK1

Impure organelle DNA: shotgun conditions apply.

Plasmid or phage DNA from hosts that exchange genes with *E. coli*

If plasmid or phage genome does not contain harmful genes or if DNA segment 99% pure and characterized   P1+EK1

Otherwise, shotgun conditions apply.

Plasmids and phage from hosts which do not exchange genes with *E. coli*

Shotgun conditions apply, unless minimal risk that recombinant will increase pathogenicity or ecological potential of the host, then   P2+EK2 or P3+EK1

NB. cDNAs synthesized *in vitro* from cellular or viral RNAs are included in above categories.

c. Animal virus vectors

Defective polyoma virus+DNA from non-pathogen	P3
Defective polyoma virus+DNA from Class 2 agent	P4
If cloned recombinant contains no harmful genes and host range of polyoma unaltered, reduce to	P3
Defective SV40+DNA from non-pathogens	P4

If inserted DNA is 99% pure segment of prokaryotic DNA lacking toxigenic genes, or a segment of eukaryotic DNA whose function has been established and which has previously been cloned in a prokaryotic host-vector system, and if infectivity of SV40 in human cells unaltered   P3

Defective SV40 lacking substantial section of the late region+DNA from non-pathogens, if no helper used and no virus particles produced   P3

Defective SV40+DNA from non-pathogen can be used to transform established lines of non-per-

missive cells under P3 provided no infectious
particles produced. Rescue of SV40 from such
cells requires   P4

d. Plant host-vector systems

P2 conditions can be approximated by insect-free greenhouses, steriliza-
tion of plant, pots, soil and runoff water, and use of standard microbiological practice.
P3 conditions require use of growth chambers under negative pressure
and routine fumigation for insect control.
Otherwise, similar conditions to those prescribed for animal systems
apply.

Appendix II:
Extracts from the Guidelines for the Use of Recombinant DNA Molecule Technology in the City of Cambridge, Massachusetts

Recommendations and Findings Submitted to the Commissioner of Health and Hospitals, December 21, 1976, and the City Manager, January 5, 1977, by the Cambridge Experimentation Review Board

Sr. Mary Lucille Banach
John L. Brush, M.D.
Daniel J. Hayes, Chairperson
Mrs. Constance Hughes

Sheldon Krimsky, Ph.D.
William J. LeMessurier
Mrs. Mary Nicoloro
Mrs. Cornelia Wheeler

Introduction

The Cambridge Experimentation Review Board (CERB) has spent nearly four months studying the controversy over the use of the recombinant DNA technology in the City of Cambridge. The following charge was issued to the board by the city manager at the request of the city council on August 6, 1976.

The broad responsibility of the Experimentation Review Board shall be to consider whether research on recombinant DNA which is proposed to be conducted at the P3 level of containment in Cambridge may have any adverse effect on public health within the city, and for this purpose to undertake, among other studies to:

(a) review the "Decision of the Director, National Institutes of Health, to Release Guidelines for Research on Recombinant DNA Molecules," dated and released on June 23, 1976;
(b) review but not be limited to the methods of physical and biological containment recommended by the NIH;
(c) review methods for monitoring compliance with applicable procedural safeguards;

    (d) review methods for monitoring compliance with
        safeguards applicable to physical containment;
    (e) review procedures for handling accidents (e.g., fire) in
        recombinant DNA research facilities;
    (f) advise the commissioner of health and hospitals on the
        reviews, findings and recommendations.

Throughout our inquiry we recognized that the controversy over recombinant DNA research involves profound philosophical issues that extend beyond the scope of our charge. The social and ethical implications of genetic research must receive the broadest possible dialogue in our society. That dialogue should address the issue of whether all knowledge is worth pursuing. It should examine whether any particular route to knowledge threatens to transgress upon our precious human liberties. It should raise the issue of technology assessment in relation to long-range hazards to our natural and social ecology. Finally, a national dialogue is needed to determine how such policy decisions are resolved in the framework of participatory democracy.

In the several months of testimony, we have come to appreciate the brilliant scientific achievements made in molecular biology and genetics. Recombinant DNA technology promises to contribute to our fundamental knowledge of life processes by providing basic understanding of the function of the gene. The benefits to be derived from this research are uncertain at this time, but the possibility for advancement in clinical medicine as well as in other fields surely exists. While we should not fear to increase our knowledge of the world, to learn more of the miracle of life, we citizens must insist that in the pursuit of knowledge appropriate safeguards be observed by institutions undertaking the research. Knowlege, whether for its own sake or for its potential benefits to humankind, cannot serve as a justification for introducing risks to the public unless an informed citizenry is willing to accept those risks. Decisions regarding the appropriate course between the risks and benefits of potentially dangerous scientific inquiry must not be adjudicated within the inner circles of the scientific establishment. Moreover, the public's awareness of scientific results that have an important impact on society should not depend on crisis situations. Many of the fears over scientific research held by the citizenry result from a lack of understanding about the nature of and the manner in which the research is conducted.

The members of CERB have made a determined effort to assess the risks to the Cambridge community of recombinant DNA research at the P3 level of physical containment. NIH, in issuing its guidelines, sought a balance between "stifling research through excessive regulation and allowing it to continue with sufficient controls." The function of CERB was not to repeat NIH's long and careful deliberation, perhaps one of the most intensive biohazards studies in the history of biology. Our role was to examine the controversy within science. We called upon people from diverse fields to testify. We encouraged skepticism, and in doing so were able to determine the locus of the controversy.

Many of us felt that it was the role of the proponents of the research to justify that *no reasonable likelihood* exists in which the public's health would be compromised if the research is undertaken under the guidelines issued by NIH. We recognized that absolute assurance was an impossible expectation. It was clearly a question of how much assurance was satisfactory to the deliberating body, and in the case of CERB, that body was comprised of citizens with no special interests in promoting the research. The uncertainty we faced was not something fabricated in our community. It was expressed most eloquently by Donald Frederickson, the director of NIH, when he issued the guidelines:

> In many instances, the views presented to us were contradictory. At present, the hazards may be guessed at, speculated about, or voted upon, but they cannot be known absolutely in the absence of firm experimental data—and unfortunately, the needed data were, more often than not, unavailable.

Our recommendations call for more assurance than was called for by the NIH guidelines. We feel that under our recommendations, a sufficient number of safeguards have been built into the research to protect the public against *any reasonable likelihood* of a biohazard. For *extremely unlikely possibilities,* we have called for additional health monitoring, whereby appropriate personnel are responsible for the detection of hazardous agents, inadvertently produced, before they are able to threaten the health of the citizens in our community.

We recognize that the controversy over the use of the recombinant DNA technology was brought to the public's attention by a small group of scientists with a deep concern for their fellow citizens and responsibility to their profession. Many of these early critics are now satisfied that the potential hazards of the research are negligible when carried out under the NIH guidelines. There are also those scientists who continue to call for more stringent control over this technology, in many instances, against the majority view of their colleagues and amidst very strained personal relations. To them we owe our gratitude for broadening the context in which the issues are being discussed. The willingness of scientists on both sides of the controversy to share their knowledge with us in our determination to arrive at a reasoned decision has been an inspiration.

CERB has spent over one hundred hours in hearing testimony and carrying out its deliberations. Our decision is as unemotional and as objective as we are capable of offering. It provides a statement of conditions and safeguards that we deem necessary for P3 recombinant DNA research to be carried out in Cambridge. The members of this citizen committee have no association with the biological research in question and no member of the board has ever had formal ties to the institutions proposing the research, with the exception of one member who has taught in unallied areas at both the institutions in question. Moreover, the city manager in selecting a group of citizens representing a cross section of the Cambridge community insured that the "empathy factor"—that is, the concern that the institutions proposing the research might lose valuable funds or that qualified researchers would leave in the event of a ban on the research—was never an issue in the deliberations.

In presenting the results of our findings we wish also to express our sincere belief that a predominantly lay citizen group can face a technical scientific matter of general and deep public concern, educate itself appropriately to the task, and reach a fair decision.

Section 1:

After reviewing the guidelines issued by the director of the National Institutes of Health (NIH) for Research Involving Recombinant DNA Molecules (issued June 23, 1976) it is the unanimous judgment of the Cambridge Experimentation Review Board that recombinant DNA research can be permitted in Cambridge provided that:

The research is undertaken with strict adherence to NIH guidelines, and in addition to those guidelines the following conditions are met:

I. Institutions proposing recombinant DNA research or proposing to use the recombinant DNA technology shall prepare a manual which contains all procedures relevant to the conduct of said research at all levels of containment and that training in appropriate safeguards and procedures for minimizing potential accidents should be mandatory for all laboratory personnel.

II. The institutional biohazards committee mandated by the NIH guidelines should be broad-based in its composition. It should include members from a variety of disciplines, representation from the bio-technicians staff and at least one community representative unaffiliated with the institution. The community representative should be approved by the Health Policy Board of the City of Cambridge.

III. All experiments undertaken at the P3 level of physical containment shall require an NIH certified host-vector system of at least an EK2 level of biological containment.

IV. Institutions undertaking recombinant DNA experiments shall perform adequate screening to insure the purity of the strain of host organisms used in the experiments and shall test organisms resulting from such experiments for their resistance to commonly used therapeutic antibiotics.

V. As part of the institution's health monitoring responsibilities it shall in good faith make every attempt, subject to the limitation of the available technology, to monitor the survival and escape of the host organism or any component thereof in the laboratory worker. This should include whatever means is available to monitor the intestinal flora of the laboratory worker.

VI. A Cambridge Biohazards Committee (CBC) be established for the purpose of overseeing all recombinant DNA research that is conducted in the City of Cambridge.

  A. The CBC shall be composed of the commissioner of public health, the chairman of the health policy board and a minimum of three members to be appointed by the city manager.

  B. Specific responsibilities of the CBC shall include:
    1. Maintaining a relationship with the institutional biohazards committees.
    2. Reviewing all proposals for recombinant DNA research to be conducted in the City of Cambridge for compliance with the current NIH guidelines.
    3. Developing a procedure for members of institutions where the research is carried on to report to the CBC violations either in technique or established policy.
    4. Reviewing reports and recommendations from local institutional biohazards committees.
    5. Carrying out site visits to insitutional facilities.
    6. Modifying these recommendations to reflect future developments in federal guidelines.
    7. Seeing that conditions designated as I-V in this section are adhered to.

Section 2:

We recommend that a city ordinance be passed to the effect that any recombinant DNA molecule experiments undertaken in the city which are not in strict adherence to the NIH guidelines as supplemented in Section 1 of this report constitute a health hazard to the City of Cambridge.

Section 3:

We urge that the city council of Cambridge, on behalf of this board and the citizenry of the country, make the following recommendations to the Congress:

I. That all uses of recombinant DNA molecule technology fall under uniform federal guidelines and that legislation be enacted in Congress to insure conformity to such guidelines in all sectors, both profit and nonprofit, whether such legislation takes a form of licensing or regulation, and that Congress appropriate sufficient funding to adequately enforce compliance with the legislation.

II. That the NIH or other agencies funding recombinant DNA research require institutions to include a health-monitoring program as part of their funding proposal and that monies be provided to carry out the monitoring.

III. That a federal registry be established of all workers participating in recombinant DNA research for the purpose of long-term epidemiological studies.

IV. That federal initiative be taken to sponsor and fund research to determine the survival and escape of the host organism in the human intestine under laboratory conditions.

ABOUT THE AUTHOR

JUNE GOODFIELD has produced and directed a number of documentary films on science, one of which won the Bronze Medal at the Venice Film Festival. Besides writing *The Siege of Cancer* and a scientific thriller (*Courier to Peking*), she is the co-author of *Architecture of Matter* (1964), and *The Discovery of Time* (1965). A Fellow of the Royal Society of Medicine, June Goodfield has taught at Michigan State University, Harvard, Wellesley, the University of Leeds and Rockefeller University. Presently she is adjunct professor at Rockefeller University and holder of a Rockefeller Foundation Humanities Fellowship.